T0301515

GUIDELINES FOR INITIATING EVENTS AND INDEPENDENT PROTECTION LAYERS IN LAYER OF PROTECTION ANALYSIS

GUIDELINES FOR INITIATING EVENTS AND INDEPENDENT PROTECTION LAYERS IN LAYER OF PROTECTION ANALYSIS

Center for Chemical Process Safety
New York, NY

Published by John Wiley & Sons, Inc., Hoboken, New Jersey.
Published simultaneously in Canada.

For general information on our other products and services or for technical support, please contact our Customer Care Department within the United States at (800) 762-2974, outside the United States at (317) 572-3993 or fax (317) 572-4002.

Wiley also publishes its books in a variety of electronic formats. Some content that appears in print may not be available in electronic formats. For more information about Wiley products, visit our web site at www.wiley.com.

Library of Congress Cataloging-in-Publication Data:

Guidelines for initiating events and independent protection layers in layer of protection analysis / Center for Chemical Process Safety of the American Institute of Chemical Engineers.
    pages cm
  Includes index.
    Summary: "Presents a brief overview of Layer of Protection Analysis (LOPA)and its variations, and summarizes terminology used for evaluating scenarios in the context of a typical incident sequence"— Provided by publisher.
  ISBN 978-0-470-34385-2 (hardback)
  1. Chemical process control—Safety measures. 2. Chemical processes—Safety measures. 3. Chemical plants—Risk assessment. I. American Institute of Chemical Engineers. Center for Chemical Process Safety.
  TP155.75.G854 2014
  660'.2815—dc23   2014012633

10 9 8 7 6 5 4

This book is one in a series of process safety guidelines and concept books published by the Center for Chemical Process Safety (CCPS). Refer to www.wiley.com/go/ccps for a full list of titles in this series.

It is sincerely hoped that the information presented in this document will lead to an even more impressive safety record for the entire industry. However, the American Institute of Chemical Engineers, its consultants, the CCPS Technical Steering Committee and Subcommittee members, their employers, their employers' officers and directors, and Process Improvement Institute, Inc., and its employees do not warrant or represent, expressly or by implication, the correctness or accuracy of the content of the information presented in this document. As between (1) American Institute of Chemical Engineers, its consultants, CCPS Technical Steering Committee and Subcommittee members, their employers, their employers' officers and directors, and Process Improvement Institute, Inc., and its employees, and (2) the user of this document, the user accepts any legal liability or responsibility whatsoever for the consequences of its use or misuse.

CONTENTS

LIST OF DATA TABLES

Initiating Events and Initiating Event Frequencies

Independent Protection Layers and Probabilities of Failure on Demand

ACRONYMS AND ABBREVIATIONS

ACGIH – American Conference of Governmental Industrial Hygienists
AIChE – American Institute of Chemical Engineers
AIHA – American Industrial Hygiene Association
ALARP – As Low As Reasonably Practicable
ALOHA – Areal Locations of Hazardous Atmospheres
ANSI – American National Standards Institute
API – American Petroleum Institute
APJ – Absolute Probability Judgment
ASME – American Society of Mechanical Engineers
ASSE – American Society of Safety Engineers
ATEX – Atmosphères Explosibles (Europe)

BEP – Best Efficiency Point
BLEVE – Boiling Liquid Expanding Vapor Explosion
BMS – Burner Management System
BPCS – Basic Process Control System
BPVC– Boiler and Pressure Vessel Code (ASME)
BS – British Standards (UK)

CCPS – Center for Chemical Process Safety (of AIChE)
CFR – Code of Federal Regulations (USA)
CPR – Committee for the Prevention of Disasters (The Netherlands)
CPQRA – Chemical Process Quantitative Risk Analysis
CPU – Central Processing Unit (Logic Solving Integrated Circuit)
CR – Contractor Technical Report (by the Nuclear Regulatory Commission, USA)
CSB – Chemical Safety Board (USA)

DCS – Distributed Control System
DDT – Deflagration-to-Detonation Transition
DIN – Deutsches Institut für Normung (Germany)

EGIG – European Gas Pipeline Incident Data Group
EPA – Environmental Protection Agency (USA)
ESD – Emergency Shutdown Device
ETA – Event Tree Analysis

FMEA – Failure Mode and Effects Analysis
FMECA – Failure Modes, Effects, and Criticality Analysis
FRP – Fiber-Reinforced Plastic
FTA – Fault Tree Analysis

GCPS – Global Congress on Process Safety (of AIChE)

HAZMAT – Hazardous Material
HAZOP – Hazard and Operability; as in HAZOP Analysis or HAZOP Study
HEART – Human Error Assessment and Reduction Technique
HEP – Human Error Probability
HERA – Human Event Repository and Analysis
HRA – Human Reliability Analysis
HCR – Human Cognitive Reliability
HMI – Human-Machine Interface

I/O – Input/Output
IE – Initiating Event
IEF – Initiating Event Frequency
IEC – International Electrotechnical Commission
IEEE – The Institute of Electrical and Electronics Engineers
IEF – Initiating Event Frequency
IPL – Independent Protection Layer
IPS – Instrumented Protective System
IRT – Independent Protection Layer (IPL) Response Time
ISA – International Society of Automation
ISO – International Organization for Standardization
ITPM – Inspection, Testing, and Preventive Maintenance

LOC – Loss of Containment
LOPA – Layer of Protection Analysis
LPG – Liquified Petroleum Gas

MAWP – Maximum Allowable Working Pressure
MOC – Management of Change
MPS – Machine Protection System
MSP – Maximum Setpoint
MSS – Manufacturers Standardization Society

NOAA – National Oceanic and Atmospheric Administration (USA)
NFPA – National Fire Protection Association
NPRD – Nonelectric Parts Reliability Data
NRC – Nuclear Regulatory Commission (USA)
NRCC – National Research Council Canada
NTSB – National Transportation Safety Board (USA)
NUREG – U.S. Nuclear Regulatory Commission Document

OREDA – Offshore Reliability Data
OSHA – Occupational Safety and Health Administration (USA)

PERD – Process Equipment Reliability Database
PES – Programmable Electronic System
PFD – Probability of Failure on Demand
PFD$_{avg}$ – Average Probability of Failure on Demand
PHA – Process Hazard Analysis
P&ID – Piping & Instrumentation Diagram
PID – Proportional–Integral–Derivative
PLT – Process Lag Time
PMI – Positive Material Identification
PPE – Personal Protective Equipment
PRV – Pressure Relief Valve
PSF – Performance Shaping Factor
PSM – Process Safety Management
PST – Process Safety Time

QRA – Quantitative Risk Assessment

RAGAGEP – Recognized and Generally Accepted Good Engineering Practice
RBPS – Risk Based Process Safety
RD – Rupture Disk
RFO – Restrictive Flow Orifice
RRF – Risk Reduction Factor

SCAI – Safety Controls, Alarms, and Interlocks
SIF – Safety Instrumented Function
SIL – Safety Integrity Level
SIS – Safety Instrumented System
SLIM – Success Likelihood Index Method
SME – Subject Matter Expert
SPAR–H – Standardized Plant Analysis Risk Model – Human Reliability Analysis
SPIDR™ – System and Part Integrated Data Resource

THERP – Technique for Human Error Rate Prediction
TR – Technical Report (by ISA)

UL – Underwriters Laboratory
USCG – United States Coast Guard

VRV – Vacuum Relief Valve
VPRV – Vacuum Pressure Relief Valve
VSV – Vacuum Safety Valve

GLOSSARY

Administrative Control	Procedural mechanism for controlling, monitoring, or auditing human performance, such as lockout/tagout procedures, bypass approval processes, car seals, and permit systems.
Asset Integrity	A risk-based process safety element involving work activities that help ensure that equipment is properly designed, installed in accordance with specifications, and remains fit for purpose over its life cycle. (Previously referred to as "mechanical integrity.")
Average Probability of Failure on Demand (PFD$_{avg}$)	Average PFD over the proof test interval of an equipment item.
Basic Process Control System (BPCS)	System that responds to input signals from the process, its associated equipment, other programmable systems and/or operator and generates output signals causing the process and its associated equipment to operate in the desired manner but that does not perform any safety instrumented functions with a claimed SIL ≥ 1 (IEC 61511 2003).
Bathtub Curve	Typical plot of equipment failure rate as a function of time. It is used to characterize the equipment lifecycle, such as early or premature failure, steady-state or normal operation failure, and wear out or end of useful life failure.
Beta Factor	A mathematical term applied in the PFD$_{AVG}$ to account for the fraction of the probability of failure that is due to dependent, or common cause, failure within the system.

Car Seal
A metal or plastic cable used to fix a valve in the open position (car sealed open) or closed position (car sealed closed). Proper authorization, controlled via administrative procedures, is obtained before operating the valve.

Chain Lock
A chain that is wrapped through or over a valve handle and locked to a support to prevent inadvertent repositioning of a valve once it is in its correct position. Removal is intended to occur only after approval is received from someone with authority and after checking that all prerequisites are met. The chain and lock provides an easy inspection aid to visually verify that the valve is in the intended position.

Clean Service
The process fluids and/or conditions do not result in fouling, corrosion, erosion, or deposition that negatively impacts the performance of a layer of protection, such as polymer formation under, in, or downstream of a relief valve.

Compensating Measures
Planned and documented methods for managing risks. They are implemented temporarily during any period of maintenance or of process operation with known faults or failures in an IPL, where there is an increased risk.

Common Cause Failure
Failure of more than one device, function, or system due to the same cause.

Common Mode Failure
A specific type of common cause failure in which the failure of more than one device, function, or system occurs due to the same cause, and failure of the devices occurs in the same manner.

Conditional Modifier
One of several possible probabilities included in scenario risk calculations, generally when the risk criteria are expressed in impact terms (e.g., fatalities) instead of loss event terms (e.g., release, loss-of-containment, vessel rupture).

Consequence
The undesirable result of an incident, usually measured in health and safety effects, environmental impacts, loss of property, and business interruption costs.

Dangerous Failure Rate
The rate (normally expressed in expected number of failures per year) that a component fails to an unsafe state/mode. (Other failure states or modes may lead to spurious trips of a system, but they do not lead to the unsafe condition of interest.)

Demand Mode	Dormant or standby operation where the IPL takes action only when a process demand occurs and is otherwise inactive. Low demand mode occurs when the process demand frequency is less than once per year. High demand mode occurs when the process demands happen more than once per year.
Dormant	A state of inactivity until a specific parametric level is reached.
Enabling Condition	Operating conditions necessary for an initiating cause to propagate into a hazardous event. Enabling conditions do not independently cause the incident, but must be present or active for it to proceed.
Event	An occurrence involving the process caused by equipment performance, human action, or external influence.
Frequency	Number of occurrences of an event per unit time (typically per year).
Human Error Probability (HEP)	The ratio between the number of human errors of a specific type and the number of opportunities for human errors on a particular task or within a defined time period. Synonyms: human failure probability and task failure probability.
Independent Protection Layer (IPL)	A device, system, or action that is capable of preventing a scenario from proceeding to the undesired consequence without being adversely affected by the initiating event or by the action of any other protection layer associated with the scenario.
Independent Protection Layer Response Time (IRT)	The IPL Response Time is the time necessary for the IPL to detect the out-of-limit condition and complete the actions necessary to stop progression of the process away from the safe state.
Incident Scenario	A hypothetical sequence of events that includes an initiating event and failure of any safeguards that ultimately results in a consequence of concern.
Initiating Event (IE)	A device failure, system failure, external event, or wrong action (or inaction) that begins a sequence of events leading to a consequence of concern.
Initiating Event Frequency (IEF)	How often the IE is expected to occur; in LOPA, the IEF is typically expressed in terms of occurrences per year.

Inspection, Testing, and Preventive Maintenance (ITPM)	Scheduled proactive maintenance activities intended to (1) assess the current condition and/or rate of degradation of equipment, (2) test the operation/functionality of the equipment, and/or (3) prevent equipment failure by restoring equipment condition. ITPM is an element of asset integrity.
Maximum Setpoint (MSP)	The maximum setpoint for an IPL is the point of maximum process deviation from the normal condition that would still allow sufficient time for the IPL to detect the deviation, to take action, and for the process to respond, preventing the consequence of concern. For SIS, this is called Maximum SIS Setpoint (MSP) per ISA-TR84.00.04 (2011).
Must	This *Guidelines* subcommittee believes that the IEF, PFD, or other aspect of an IE or IPL is valid only if the listed criteria are met. "Must" can also be used in reference to basic definitions.
Passive Fluid	Nonreactive and nonhazardous fluid.
Performance Shaping Factors (PSF)	Factors that influence the likelihood of human error.
Probability of Failure on Demand (PFD)	The likelihood that a system will fail to perform a specified function when it is needed.
Process Lag Time (PLT)	The process lag time indicates how much time it will take for the process to respond and avoid the consequence of concern, once the IPL has completed its action.
Process Safety Time (PST)	The time period between a failure occurring in the process, or its control system, and the occurrence of the consequence of concern.
Risk	A measure of potential economic loss, human injury, or environmental impact in terms of the frequency of the loss or injury occurring and the magnitude of the loss or injury if it occurs.
Safeguard	Any device, system, or action that either interrupts the chain of events following an initiating event or that mitigates the consequences. *Not all safeguards will meet the requirements of an IPL.*

Safety Controls, Alarms, and Interlocks (SCAI)	Process safety safeguards implemented with instrumentation and controls, used to achieve or maintain a safe state for a process, and required to provide risk reduction with respect to a specific hazardous event (ANSI/ISA 84.91.01 2012). These are sometimes called safety critical devices or critical safety devices.
Safety Instrumented Function (SIF)	A safety function allocated to a Safety Instrumented System (SIS) with a Safety Integrity Level (SIL) necessary to achieve the required risk reduction for an identified scenario of concern.
Safety Integrity Level (SIL)	One of four discrete ranges used to benchmark the integrity of each SIF and the SIS, where SIL 4 is the highest and SIL 1 is the lowest.
Safety Instrumented System (SIS)	A separate and independent combination of sensors, logic solvers, final elements, and support systems that are designed and managed to achieve a specified Safety Integrity Level (SIL). A SIS may implement one or more Safety Instrumented Functions (SIFs).
Severity	A measure of the degree of impact of a particular consequence.
Should	This *Guidelines* subcommittee believes that an alternative protocol to achieve the same criteria/goal is acceptable.
Systematic Error	Also referred to as "systemic error." ISA-TR84.00.02 (2002) defines systematic error as "an error that occurred during the specification, design, implementation, commissioning, or maintenance."
Validation	Activity of demonstrating that the installed equipment and/or associated human actions achieve the core attributes and the requirements of the design basis. Testing is one approach to validation.
Verification	Activity of making sure the equipment is installed to specification. (In the case of a Safety Instrumented Function (SIF), SIL verification often refers to calculating the PFD_{avg} of a SIS to ensure that it achieves the stipulated SIL.)

ACKNOWLEDGMENTS

The American Institute of Chemical Engineers (AIChE) and the Center for Chemical Process Safety (CCPS) express their appreciation and gratitude to the members of the *Guidelines in Initiating Events and Independent Protection Layers in Layer of Protection Analysis* subcommittee of the CCPS Technical Steering Committee for providing input, reviews, technical guidance, and encouragement to the project team throughout the preparation of this book. CCPS expresses gratitude to the team member companies for their generous support of this effort. CCPS also expresses appreciation to the members of the Technical Steering Committee for their advice and support in the writing of this book.

Subcommittee Members for *Guidelines for Initiating Events and Independent Protection Layers in Layer of Protection Analysis.* CCPS thanks the *Guidelines for Initiating Events and Independent Protection Layers in Layer of Protection Analysis* subcommittee for their significant efforts and their contributions to advancing the practice of LOPA. Subcommittee members included:

Wayne Chastain, *Chair*	Eastman Chemical Company
John Baik	BP
Matt Bennett	BP
Tony Clark	Process Improvement Institute, Inc.
Jim Curtis	Celanese
Rick Curtis	ABS Consulting
Tom Dileo	Albemarle
Richard R. Dunn	E.I. Du Pont de Nemours & Company, Inc.
Randy Freeman	S&PP Consulting
Bob Gale	Emerson Process Management
Kathleen A. Kas	The Dow Chemical Company
Kelly Keim	ExxonMobil Chemical Company
Kevin Klein	Celanese
Don Lorenzo	ABS Consulting
Steve Meszaros	Wyeth
John Remy	Lyondell Basell

Angela Summers	SIS-TECH
Scott Swanson	Intel Corporation
Hal Thomas	Air Products (later, Exida)
Stanley Urbanik	E.I. Du Pont de Nemours & Company, Inc.
Tim Wagner	The Dow Chemical Company
Scott Wallace	Olin Corporation
Robert Wasileski	NOVA Chemicals, Inc.
Paula Wiley	Chevron Phillips Chemical Company LP

CCPS thanks Bill Bridges and the Process Improvement Institute (PII), who prepared the initial the peer review manuscript on behalf of the subcommittee. Wayne Chastain and Kathy Kas led the revision of the peer review document into the final consensus version published herein. The efforts of Sheila Vogtmann (SIS-TECH) in editing the final text were also much appreciated.

The CCPS Staff Consultant was John F. Murphy, who coordinated meetings and facilitated subcommittee reviews and communications.

Peer Reviewers for *Guidelines for Initiating Events and Independent Protection Layers in Layer of Protection Analysis*

Before publication, all CCPS books are subjected to a thorough peer review process. CCPS gratefully acknowledges the thoughtful comments and suggestions of the peer reviewers. Their work enhanced the accuracy and clarity of this book. Although the peer reviewers have provided many constructive comments and suggestions, they were not asked to endorse this book and were not shown the final draft before its release.

Joe Allaben	Flint Hills Resources
John Alderman	Aon Energy Risk Engineering
Mohamad Fazaly Mohamad Ali	Petronas
Christie Arseneau	Momentive Specialty Chemicals Inc.
Brian Baer	Brian Baer
Kepa Bengoetxea	The Dow Chemical Company
Kumar Bhimavarapu	FM Global
Christine E. Browning	Eastman Chemical Company
Art Dowell, III	Retired from Rohm and Haas Company/Dow Chemical Company
Dale E. Dressel	Solutia Inc.
Richard Gowland	European Process Safety Centre
Sara B. Guler	The Dow Chemical Company
Robert W. Johnson	Unwin Company
Leonard Laskowski	Emerson Process Management
David K. Lewis	NOVA Chemicals, Inc.
David Lewis	Occidental Chemical Corporation

Pete Lodal	Eastman Chemical Company
Keith R. Pace	Praxair, Inc.
Paul Delanoy	The Dow Chemical Company
Hasim Sakarya	The Dow Chemical Company
Irfan Shaikh	Scandpower Inc./Lloyd's Register
Adrian L. Sepeda	Sepeda Consulting
Karen Shaw Study	The Dow Chemical Company
Laurence Thring	Huntsman Holland B.V.
Florine W. Vincik	Syngenta
Harry White	Styron, LLC
Ronald J. Willey	Northeastern University
John A. Williamson	Flint Hills Resources
John C. Wincek	Croda Inc
Klaus Wischnewski	DuPont Performance Coatings GmbH

PREFACE

The American Institute of Chemical Engineers (AIChE) has been closely involved with process safety and loss control issues in the chemical and allied industries for more than four decades. Through its strong ties with process designers, constructors, operators, safety professionals, and members of academia, AIChE has enhanced communication and fostered continuous improvement of the industry's high safety standards. AIChE publications and symposia have become information resources for those devoted to understanding the causes of incidents and discovering better means of preventing their occurrence and mitigating their consequences.

The Center for Chemical Process Safety (CCPS) was established in 1985 by AIChE to develop and disseminate technical information for use in the prevention of major chemical incidents. CCPS is supported by more than 140 sponsoring companies in the chemical process industry and allied industries; these companies provide the necessary funding and professional experience for its technical subcommittees.

The first CCPS project was the preparation of *Guidelines for Hazard Evaluation Procedures* (CCPS 1985). CCPS achieved its stated goal with the publication of this book in 1985 and has since continued to foster the development of process safety professionals in all industries. For example, CCPS has developed more than 100 Guidelines and Concept Books and has sponsored numerous international meetings since its inception. A number of other projects are ongoing. This activity has occurred in the midst of many other changes and events that, throughout the past years, have fostered an unprecedented interest in hazard evaluation.

Layer of protection analysis (LOPA) is a streamlined tool for analyzing and assessing risk. LOPA has grown in popularity in the last decade since the publication of the first CCPS/AIChE Concept Book on the subject, *Layer of Protection Analysis: Simplified Process Risk Assessment (CCPS LOPA)*, (CCPS 2001). LOPA generally uses order-of-magnitude estimates of frequency, probability, and consequence severity, together with conservative rules. This book builds on *CCPS LOPA* (2001) by providing additional examples of initiating events (IE) and independent protection layers (IPL). More complete guidance is offered on how to determine the value of each prospective IE frequency and IPL PFD. Finally, there is more elaboration on the management systems that an organization should have in place to qualify an IE or IPL at a given value.

As is true for other CCPS books, this document does not contain a complete program for managing the risk of chemical operations, nor does it give specific advice on how to establish a risk analysis program for a facility or an organization. However, it does provide insights that should be considered when performing more detailed, scenario-based risk evaluations.

Guidance in this document cannot replace hazard evaluation experience. This document should be used as an aid for the further training of hazard analysts and as reference material for experienced practitioners. Only through both study and experience will hazard analysts become skilled in the identification of initiating events and independent protection layers. Using this document within the framework of a complete process safety management (PSM) program will help organizations continually improve the safety of their facilities and operations.

1

INTRODUCTION

Layer of protection analysis (LOPA) is a simplified quantitative tool for analyzing and assessing risk. LOPA was developed by user organizations during the 1990s as a streamlined risk assessment tool, using conservative rules and order-of-magnitude estimates of frequency, probability, and consequence severity. When the method was shown to be an efficient means to assess risk, several companies published papers describing the driving forces behind their efforts to develop the method, their experience with LOPA, and examples of its use. In particular, the papers and discussion among the attendees at the Center for Chemical Process Safety (CCPS) International Conference and Workshop on Risk Analysis in Process Safety in 1997 brought agreement that a book describing the LOPA method should be developed. This led to the publication of the Concept Book *Layer of Protection Analysis: Simplified Process Risk Assessment (CCPS LOPA)* in 2001. Since its inception, the LOPA methodology has continued to evolve, and some companies have utilized or supplemented the methodology with more advanced techniques.

LOPA has grown greatly in popularity and usefulness since the publication of *CCPS LOPA* (2001) on the subject. *Guidelines for Initiating Events and Independent Protection Layers in Layer of Protection Analysis* builds on LOPA by

- Providing additional examples of initiating events (IEs) and independent protection layers (IPLs)
- Providing more guidance for determining the value of each prospective initiating event frequency (IEF) and IPL probability of failure on demand (PFD)
- Providing more information on the overall management systems, as well as other considerations specific to a particular IE or IPL, which are needed to support the use of the values provided in this document

This chapter will

- Identify the audience for this book
- Define the scope of this book
- Describe differences between this book and *CCPS LOPA* (2001)
- Recap the LOPA approach and provides a brief description of how the method has evolved since its inception
- Discuss linkages between this book and other publications

1.1   AUDIENCE

This book is intended for the following readers:

- Current practitioners of LOPA. It is assumed that readers of this book will have read, understood, and applied the principles of *CCPS LOPA* (2001). These practitioners may include process engineers, risk analysts, and process safety specialists. For this audience, Chapters 3 through 5 provide more information regarding the application of LOPA and additional examples of IEs and IPLs. Chapter 6 and the appendices contain guidance for analysts who want to supplement the LOPA approach with more detailed methods, such as fault tree analysis (FTA), event tree analysis (ETA), and human reliability analysis (HRA).

- Executives who are considering expanding their corporate strategy for managing risk by adding LOPA to their existing risk analysis process. For the executive audience, Chapter 2 discusses key elements of LOPA and the management systems needed to support claims regarding IEF and IPL PFD values.

- Project managers who want to ensure that a new process or process modification has sufficient layers of protection. LOPA is a tool for selecting and evaluating alternative design options and layers of protection that can be used during any phase of a capital project.

- Engineers, chemists, operations and maintenance personnel, supervisors, department managers, and others who ensure that the technical and administrative requirements for each IE and IPL are met. The intent is to ensure that IEs occur at or below their assumed frequencies and that IPLs perform at least as well as their assumed PFDs. A major goal of this book is to reinforce the activities and documentation that are needed to achieve and maintain the estimates provided for each IEF and IPL PFD value. Chapters 3 through 6 are useful for this audience.

1.2 SCOPE

CCPS LOPA (2001) set the guidelines for using LOPA as an intermediate method between qualitative risk analysis typically used to support risk ranking following hazard evaluation/analysis and quantitative analysis as described in *Guidelines for Chemical Process Quantitative Risk Analysis* (CCPS 2000). This book builds on the foundation laid by *CCPS LOPA* (2001) by clarifying key concepts and reinforcing the limitations and the requirements of this method. The main scope of this book is to provide examples of IEs and IPLs and useful guidance on the activities and documentation needed to achieve and maintain the IEF and IPL PFD values suggested in this text. This guidance is intended to help companies or facilities develop appropriate values for their own LOPA protocols.

This document is not a second edition of *CCPS LOPA* (2001) and is not intended to change the basic criteria established for LOPA in that text. Rather, this document reinforces the basic criteria in *CCPS LOPA* (2001), providing examples where appropriate. However, since the publication of *CCPS LOPA* (2001), industry has developed further knowledge and understanding of the methodology through experience. Additional IEs and IPLs have been proposed, and some previously suggested IEF and IPL PFD values have been changed. As a result, practitioners have requested more details regarding the identification and application of IEs and IPLs in LOPA. CCPS has, therefore, seen the need to provide additional guidance for selecting an IE or IPL PFD value.

This document forgoes detailed explanations of the analysis and design requirements for safety controls, alarms, and interlocks (SCAI), since these requirements are addressed in other CCPS publications, such as *Guidelines for Safe Automation of Chemical Processes* (CCPS 1993) and *Guidelines for Safe and Reliable Instrumented Protective Systems* (CCPS 2007b), and in industry standards, such as IEC 61511 (2003) and ANSI/ISA 18.2 (2009). This document does provide guidance on the risk reduction that can be claimed in a Basic Process Control System (BPCS) or a Safety Instrumented System (SIS) and recommends specific design and management practices to support these claims. This document does not discuss the design of SCAI or other required activities to ensure the reliability of SCAI throughout their lifecycle. The reader should refer to the appropriate industry standards to ensure that the implemented instrumented system complies with good engineering practice.

This document excludes detailed explanations of conditional modifiers, which are probability factors used to estimate the likelihood of fires, explosions, and fatalities once a release has occurred. A limited amount of guidance was provided in *CCPS LOPA* (2001), and the topic is treated in more depth in the recent *Guidelines for Enabling Conditions and Conditional Modifiers in Layer of Protection Analysis* (CCPS 2013).

1.3   KEY CHANGES SINCE THE INITIAL *LOPA* CONCEPT BOOK

The initial *CCPS LOPA* (2001) book established the concept of layers of protection analysis and how it could be used to estimate risk. Since that time, LOPA concepts have been incorporated into a wide range of techniques, from simple scenario-based analysis of the potential for loss events to complex cumulative risk analysis of the potential for specific types of harm. LOPA has been applied throughout the process industry sector and has influenced the practice of risk analysis worldwide. After 12 years of use, the LOPA approach continues to evolve, and additional guidance has been provided. Below is a summary of the key changes from *CCPS LOPA* (2001).

The first significant change is in the more detailed discussion of the individual IEs and IPLs that are included in the text. Some values for IEFs and IPL PFDs were provided in *CCPS LOPA* (2001). Guidance was provided in general terms regarding the appropriate selection of IEFs and the underlying assumptions associated with the values selected. For IPLs, the basic requirements of independence, effectiveness, and auditability were discussed and some PFDs were provided; however, these topics were also covered in general terms. This book provides a data table for each IE and IPL that not only gives a suggested IEF or IPL PFD value but also provides the recommended design, operation, maintenance, and testing guidance associated with that value.

A second significant change is in the treatment of pressure relief systems and the differentiation in this document among different relief applications. General guidance on the use of relief valves and rupture disks as IPLs was provided in *CCPS LOPA* (2001). In this document, suggested values have been provided for relief systems of various types. The document also highlights the importance of having a strong management system in place to ensure that valves that can isolate relief devices from the process are maintained in an open position.

CCPS LOPA (2001) discussed the need for an IPL to be independent of other IPLs and of the initiating event. Common cause failure can occur when there is the potential for the failure of more than one component or system as a result of a single failure. The *Guidelines* subcommittee considered the potential for common cause failure when developing generic IPL PFD values for certain IPLs in Chapter 5. In the case of dual pressure relief valves in series, the suggested generic PFD valve for the combined system was adjusted to account for the potential for common cause failure due to the likelihood that both IPLs would be similar devices, maintained at the same time, and exposed to identical process conditions.

At the time of *CCPS LOPA* (2001), check valves were not generally considered to be valid IPLs due to a lack of data supporting their reliability. Since that time, understanding of check valve reliability has improved, assisted by more data that substantiates their reliability. (Refer to Appendix D for Example

Reliability Data Conversion for Check Valves for more information.) Based on this data, check valves have been included as IPLs in this book

To determine whether an IPL will be effective, it is necessary to consider the timeline of a scenario. *CCPS LOPA* (2001) recognized that, to be effective, an IPL needs to have sufficient time to take action. However, the timeline of the progression of a scenario was not discussed in detail. It is important to understand how quickly a process deviation will be detected, how much time will be required to diagnose the situation, how rapidly the IPL will be able to act, and how much time will be required by the process to respond to the IPL action. This consideration of time dependency is discussed in detail in Chapter 3.

In LOPA, it is assumed that IPLs are challenged infrequently. However, as recognized in *CCPS LOPA* (2001), there are situations where IPLs are challenged frequently. This is referred to as high demand mode. In *CCPS LOPA* (2001), an IPL was considered to be in high demand mode if it was challenged at a frequency of more than twice the proof test interval. However, recent guidance (IEC 61508-4 Section 3.5.16) (2010) has redefined high demand mode as occurring when the IPL is challenged more often than once a year. This change has been reflected in Chapter 3.

CCPS LOPA (2001) Approach B allowed two orders of magnitude credit to be claimed on a BPCS logic solver if the BPCS was not the IE. However, it cautioned that IEC 61511 (2003) was moving toward publication and that future developments might affect this allowance. *CCPS LOPA* (2001) based its position on the assumption that the BPCS central processing unit (CPU) had at least two orders of magnitude better performance than the field devices. More recent field data demonstrate that the failure rate of a typical BPCS CPU is greater than 0.01/yr (*PDS Data Handbook* [SINTEF 2010]), so the BPCS CPU failure rate is comparable to many other electrical, mechanical, or programmable electronic devices. Section 5.2.2.1 discusses BPCS claims in more detail. This book addresses the requirements for claiming credit for two BPCS loops that share a single logic solver, whether as an IE and IPL or two IPLs. When claiming two orders of magnitude from a single controller, IEC 61511 (2003) requires that the system be designed and managed as a SIS.

In addition to addressing the number of credits that can be claimed in a BPCS controller, this document also adopted the new International Society of Automation (ISA) terminology issued in ANSI/ISA 84.91.01 (2012). In particular, the term "Safety Controls, Alarms, and Interlocks" (SCAI) has replaced "Instrumented Protective Systems" (IPS). The term "safety instrumented system (SIS) loop" is used to define the equipment necessary to execute identified safety instrumented functions (SIF). The analysis of the system is also emphasized in the estimation of the IE frequency and the IPL PFD provided by specific BPCS or SIS architectures. System analysis is critical to ensure that common cause and systematic failure potential is properly considered. Shared (or similar) equipment, procedures, and personnel increase this failure potential, so it is recommended that sources for common cause and systematic failure be carefully considered when assessing the performance of the instrumented systems listed for a hazard scenario.

In this document, additional guidance is provided in Appendix A regarding human error as it relates to IEs and IPLs. Validation of human error rates used in LOPA is also discussed in Chapter 2 and Appendix B, and example approaches for validation are provided.

1.4   RECAP OF LOPA

1.4.1   What Is LOPA?

LOPA is a simplified form of risk assessment, where risk is defined as a function of both the frequency and the consequence of an incident scenario. LOPA was originally developed as a streamlined risk assessment tool, using conservative rules and order of magnitude estimates of frequency, probability, and consequence severity. For consistency, this book will continue to use the order of magnitude convention of *CCPS LOPA* (2001).

The LOPA methodology does not identify potential scenarios of concern. Rather, LOPA is an analysis tool that typically builds on the information developed during a qualitative hazard evaluation, such as a process hazard analysis (PHA) or Hazard and Operability (HAZOP) Study. Chapter 2 provides some guidance on considerations for scenario selection.

LOPA was originally developed to examine the selected scenarios by focusing on how individual causes propagate to a scenario of concern. Relative to quantitative risk assessment, a LOPA scenario may represent one path through an event tree.

LOPA scenario = Simplified Risk Assessment of One Cause-Consequence Pair

Over time, the key elements of LOPA have been applied to a wide range of process applications and risk studies. Many companies use LOPA to assess the frequency of loss events. These companies focus on selecting IPLs that act to stop the event propagation before a release occurs. When calculating the frequency for a loss event scenario posing a consequence of concern, the IEF is multiplied by the product of the IPL PFDs:

$$f_i^C = IEF_i \; x \; PFD_{i1} \; x \; PFD_{i2} \; x \; ... \; x \; PFD_{ij}$$

where

f_i^C = Frequency of the consequence occurring for scenario i

IEF_i = Frequency of the IE for scenario i

PFD_{ij} = Probability of failure on demand of independent protection layer j for scenario i

Other organizations use LOPA to assess the frequency of specific types of harm posed by a scenario of concern. Companies may evaluate the frequency at which certain flammable or toxic release scenarios could affect a population. These studies incorporate *conditional modifiers*, which are factors used to determine the frequency that a specific type of harm will occur as a result of a loss event. This frequency is determined by multiplying the IEF by the product of the IPL PFDs and the probability of the various conditions that were considered in assessing the harm or consequence outcome. For example, to evaluate the scenario of a fire occurring as a result of a flammable release, the probability of ignition $P^{ignition}$ is applied, and the frequency of a fire is determined as

$$f_i^{fire} = IEF_i \times PFD_{i1} \times PFD_{i2} \times \dots \times PFD_{ij} \times P^{ignition}$$

where

f_i^{fire} is the frequency of a fire for initiating event i

$P^{ignition}$ is the probability of ignition for a flammable release

Conditional modifiers are not addressed by this document because the focus of this book is the selection of appropriate IEF and IPL PFD values in LOPA. A more detailed treatment of conditional modifiers is provided in *Guidelines for Enabling Conditions and Conditional Modifiers in Layer of Protection Analysis* (CCPS 2013) and *Guidelines for Determining the Probability of Ignition of a Released Flammable Mass* (CCPS 2013b).

Many organizations use the LOPA approach to evaluate single scenarios to ensure that no scenario exceeds a particular frequency. Others have geographic or personal risk criteria, so it is necessary to sum the frequencies of the scenarios that affect the geographic area under review. For example, the frequency of a release posing a specific consequence of concern, or the frequency at which a population is affected by loss events on a facility, is determined by

$$f^C = f_1^C + f_2^C + \dots + f_i^C$$

where

f_i^C is the frequency of the consequence of the ith initiating event

For more information regarding developing risk criteria, refer to *Guidelines for Developing Quantitative Safety Risk Criteria*, 2nd Edition (CCPS 2009a).

Some companies no longer restrict LOPA analyses to order of magnitude IEF and IPL PFD values. For example, there are companies that consider safeguards that do not provide an order of magnitude risk reduction but are expected to provide at least half of an order of magnitude risk reduction. Some organizations do not restrict the analysis to order of magnitude values where detailed quantitative analysis has been performed for the IEF or IPL PFD. Others use site-specific data when available. Regardless of the LOPA approach used, it is important that the methodology is aligned with the set of criteria used by the organization to evaluate risk.

Like other risk analysis methods, the primary purpose of LOPA is to determine whether there are sufficient layers of protection to reduce the risk below specified risk tolerance criteria. A scenario may require one or more protection layers, depending on the scenario frequency and the potential consequence severity. For any given scenario, only one layer is required to work successfully for the consequence being analyzed to be prevented. However, since no layer is 100% effective or reliable, multiple layers of protection may be necessary to lower the risk below the specified risk tolerance criteria.

1.4.2   Common Elements of LOPA

While some companies have modified the definition and approach defined in *CCPS LOPA* (2001), LOPA approaches generally share the following common features:

- A means to assess or estimate risk, which can be applied throughout the organization
- Risk tolerance criteria. Individual companies use different criteria; these criteria may be based on the potential incident frequency and consequence severity of a single scenario or of a group of scenarios.
- Criteria for crediting safeguards as IPLs
- Default values for IEFs and for the IPL PFDs
- A procedure for performing the required calculations
- A procedure for determining whether the risk associated with a scenario meets the risk tolerance criteria for an organization and, if it does not, how the risk should be managed

1.4.3   When to Use LOPA

Potential incident scenarios are typically identified by a team using a hazard evaluation method, such as those described in *Guidelines for Hazard Evaluation Procedures*, 3rd Edition (CCPS 2008a). Once incident scenarios are identified, the risks associated with these scenarios are evaluated. Qualitative risk assessment methods are often used as part of PHA to assess the frequency and consequence severity; in many cases, these methods are sufficient to understand and address the hazards. However, for some process safety hazards, qualitative methods do not provide a sufficiently detailed understanding of risk. LOPA is a method that an analyst can use to provide an order of magnitude estimate of the risk posed by an identified scenario of concern. LOPA can also be used to compare risks within a given organization.

LOPA can be a useful tool to assess and manage risk during any phase in the lifecycle of a process. During process development, LOPA can be used to evaluate and compare the risks associated with various process technology options. As the design matures, the understanding of the risk of a process also develops, and this knowledge can be applied using LOPA to assess risks associated with facility siting, process design, and the plant operating philosophy. LOPA is recognized in IEC 61511 (2003) as one of the methods that may be used to select the safety integrity level (SIL) for a SIS. After a plant process is started up, any changes to the process should be carefully evaluated and managed. LOPA can be used to help make risk judgments involving plant modifications and procedural changes during routine operations. LOPA can be applied to nearly any risk decision, from the assessment of a single process modification during a management of change review to the evaluation of major accident scenarios, such as safety cases. Refer to *CCPS LOPA* (2001) for details of when and how to use LOPA over the lifecycle of the process.

LOPA does not work well for all scenarios because of its conservative rules and order of magnitude estimates. LOPA may not be the appropriate methodology if there is a lack of independence between IE and IPLs. It can be challenging to apply LOPA to scenarios where the IE is due to human error and risk reduction is dependent primarily on procedural and administrative practices. For example, LOPA may not be as useful for scenarios related to transient modes of operation, such as start-up, shutdown, and maintenance modes, which often depend heavily on equipment design, operating discipline, and administrative controls.

LOPA should not, in general, be used as a justification for not installing standard recognized and generally accepted good engineering practice (RAGAGEP) safeguards solely because they do not qualify as an IPL or because the LOPA indicates that the facility's risk tolerance criteria are met without the safeguards. There are times when it makes sense to retain safeguards that do not meet the full order of magnitude risk reduction criteria for an IPL, because they may still provide some risk reduction or because the safeguards may provide compensating measures that are reliable for a short period of time (such as when an IPL is bypassed).

There are times when LOPA can help determine that certain safeguards are not value-added and should be deleted. One example may be the elimination of unnecessary alarms that would otherwise distract an operator from more critical alarms. The use of LOPA to justify removing a safeguard or IPL should be done cautiously. A safeguard may still have value even though a full order of magnitude decrease in risk is not required to meet the risk tolerance criteria. Some countries expect the application of the ALARP principle, which relies on the demonstration that the risk has been reduced as low as reasonably practicable (ALARP).

1.4.4   Inherently Safer Processes and LOPA

Inherently safer approaches to process design are used to eliminate or minimize a hazard. This reduces the need for safeguards and layers of protection that would otherwise be required to manage the risk of a process. Inherent safety reviews can be done prior to LOPA so that incident scenarios can be eliminated or their consequences reduced. However, LOPA can also be used to identify scenarios where use of an inherently safer approach could result in a significant risk reduction.

1.4.5   Advanced LOPA Techniques

LOPA was originally developed as a simplified quantitative risk assessment method to be used after scenarios of concern are identified. It was developed because FTA and HRA can be overly complex methods for many decisions; these methods also require specialized resources and tools to yield meaningful results. In contrast, LOPA is a technique that can be used and understood by a broader group of risk analysts. However, because of its simplified approach, the LOPA methodology does have limitations. As companies became more experienced with the LOPA method, they advanced the understanding and the use of the LOPA methodology by combining LOPA with more detailed quantitative risk assessment techniques.

This document provides guidance on the basic tenets of LOPA and discusses where more quantitative risk assessment techniques can be combined with the LOPA methodology. It is a responsibility of an organization to define their risk assessment protocols and ensure that the risk methodologies used are aligned with the organization's risk criteria. Chapter 6 provides guidance on when combining more detailed quantitative risk assessment techniques with LOPA may be appropriate. Chapter 6 also gives some examples of how these techniques may be used in conjunction with the LOPA approach.

1.5   DISCLAIMER

This document presents numerical values related to reduction of risk of catastrophic events. It also includes the related activities, practices, and analysis needed to support suggested IEF and IPL PFD values. However, the authors of this document and CCPS/AIChE cannot predict all possible applications where these generic values, and the related practices to achieve these values, might be used. Therefore, each organization that uses the guidance in this document is expected to determine values that are applicable for its situation and follow appropriate practices to maintain those values.

Guidance and values used in this document only apply to technology with which this *Guidelines* subcommittee has experience. The reliability of newer technology should be proven and have a documented history of performance. For example, some companies have begun using wireless transmission of signals from field sensors, rather than hardwired signal transmission, for process monitoring. In this case, the *Guidelines* subcommittee could not find sufficient history of performance to allow an estimate of the failure rate or probability of failure on demand of wireless systems.

1.6   LINKAGE TO OTHER CCPS PUBLICATIONS

CCPS has published many books dealing with process safety issues in the chemical industry. *Guidelines for Initiating Events and Independent Protection Layers in Layer of Protection Analysis* references techniques that are described in the following CCPS publications. In addition to these CCPS publications, a number of pertinent industry references are listed at the end of each chapter of this document.

Guidelines for Initiating Events and Independent Protection Layers in Layer of Protection Analysis has been written to be a companion to *Layer of Protection Analysis* (CCPS 2001) Concept Book, referenced as *CCPS LOPA* hereafter. This book provides recommended IEF and IPL PFD values and gives considerations for their use in LOPA.

Guidelines for Safe and Reliable Instrumented Protective Systems (CCPS 2007b) describes the instrumented protective systems and the associated management systems that can be used to reduce process risk.

Guidelines for Enabling Conditions and Conditional Modifiers in Layer of Protection Analysis (CCPS 2013) expands on these two factors that were introduced in *CCPS LOPA* (2001). The text provides guidance on the types and appropriate use of these factors as well as indications of potential misuse that could adversely affect risk analyses.

Guidelines for Determining the Probability of Ignition of a Released Flammable Mass (CCPS 2013b) further expands on one conditional modifier, the probability of ignition given a flammable release. The book provides information and methods that can be used to estimate the probability of ignition for flammable vapor and liquid releases. Software tools are also provided to calculate the ignition probability.

Guidelines for Risk Based Process Safety (CCPS 2007a), abbreviated as *RBPS* hereafter, provides updated administrative approaches and tools that help companies build and operate more effective process safety systems. *RBPS* provides guidance on how to design a process safety management system, correct a deficient system, or continuously improve an existing system. One of the pillars of *RBPS* is to "understand hazards and risk." LOPA is one technique that can be used to understand risk.

Guidelines for Hazard Evaluation Procedures, 3rd Edition (CCPS 2008a) describes methods to identify and assess the significance of hazards found in processes during all modes of operation. Key inputs to LOPA are the scenarios obtained from hazard identification and evaluation teams. LOPA is also described as a risk analysis method in this text.

LOPA is a streamlined approach for analyzing and assessing risk. It can be viewed as a simplification of the quantitative risk analysis methods described in *Guidelines for Chemical Process Quantitative Risk Analysis,* 2nd Edition (CCPS 2000).

Guidelines for Developing Quantitative Safety Risk Criteria (CCPS 2009a) describes various risk criteria that can be used in LOPA and other quantitative risk analysis methods. Although the primary focus of *Guidelines for Developing Quantitative Safety Risk Criteria* is on individual and societal risk criteria for use in chemical process quantitative risk analysis, one section of the book is dedicated to apportioning risk criteria to individual scenarios.

Inherently Safer Chemical Processes, A Life Cycle Approach, 2nd Edition (CCPS 2009b) defines the concept of inherent safety and describes the use of inherently safer approaches for hazard and risk reduction. This approach can be used to eliminate potential causes of events, reduce the frequency of scenarios, and lessen the severity of potential consequences.

Guidelines for Improving Plant Reliability through Data Collection and Analysis (CCPS 1998a) describes how equipment failure rate data and other plant reliability data can be collected and used.

Guidelines for Process Equipment Reliability Data, with Data Tables (CCPS 1989) contains numerous descriptions and references to generic equipment failure data sources as well as an introduction to equipment taxonomies.

Guidelines for Mechanical Integrity Systems (CCPS 2006) describes or references many of the best practices for maintaining equipment features in a process plant, including instrumentation, rotating equipment, fixed equipment, and support structures. This book is an excellent resource, although it does not contain values for IE frequencies or IPL PFDs. *Guidelines for Initiating Events and Independent Protection Layers in Layer of Protection Analysis* relates the appropriate maintenance activities to suggested performance values for a number of process components.

Human Factors Methods for Improving Performance in the Process Industries (CCPS 2007c) provides guidance on techniques and tools for addressing human factors. The practices described in the text support the reliability of human systems as initiating events and as independent protection layers. This book also has an extensive checklist of design and operations issues that can be addressed to improve the performance of workers.

Guidelines for the Management of Change for Process Safety (CCPS 2008b) provides guidance on the implementation of management of change (MOC) procedures. The text provides information on the initial design of an MOC system, the scope of the system throughout the plant lifecycle, and integration with other aspects of process safety management. MOC is a critical process for maintaining the reliability of independent protection layers and the managing the failure rates for initiating events used in LOPA.

1.7   ANNOTATED DESCRIPTION OF CHAPTERS

Following is a list of the chapters and appendices contained in this book and a description of the contents.

Chapter 2 provides a discussion of the management systems that should function effectively to support the values for IEFs and IPL PFDs that are provided in this book. It also reviews key aspects of the LOPA method and important elements to consider when selecting the appropriate incident scenarios, estimating the scenario IEF, assessing the potential consequences of a scenario, and applying IPLs to manage the risk.

Chapter 3 explains the core attributes of IPLs, which can also influence the frequency of initiating events. Understanding of these concepts is critical to properly apply the data tables provided in later chapters.

Chapter 4 discusses various IEs and their IEFs based on generic industry data. This chapter also gives guidance on the practices needed to achieve and maintain the failure rate values provided.

Chapter 5 discusses various IPLs and their PFDs based on generic industry data. This chapter also gives guidance on the practices needed to achieve and maintain the failure probability values provided.

Chapter 6 provides guidance on when other quantitative methods, such as FTA, ETA, and HRA, may be appropriate to use in addition to, or instead of, LOPA.

Appendices include further elaboration on some of the complex aspects of IEs and IPLs listed in the core chapters of the document.

Appendix A provides an introduction to the concepts of human factors that are used in HRA, including the performance shaping factors (PSFs) that influence human reliability.

Appendix B discusses considerations for gathering site-specific human performance data.

Appendix C explains how to derive IEFs and IPL PFDs from site data, generic published industry data, prediction methods, and expert opinion.

Appendix D provides example reliability data for check valves. Check valve design has improved over the years, and data exist that indicate that check valves can be effective IPLs when properly specified and maintained.

Appendix E provides examples of how some companies define overpressure consequences for LOPA.

REFERENCES

ANSI/ISA (American National Standards Institute/International Society of Automation). 2009. *Management of Alarm Systems for the Process Industries.* ANSI/ISA-18.2-2009. Research Triangle Park, NC: ANSI/ISA.

ANSI/ISA. 2012. *Identification and Mechanical Integrity of Safety Controls, Alarms and Interlocks in the Process Industry.* ANSI/ISA-84.91.01-2012. Research Triangle Park, NC: ANSI/ISA.

CCPS. 1989. *Guidelines for Process Equipment Reliability Data, with Data Tables.* New York: AIChE.

CCPS. 1993. *Guidelines for Safe Automation of Chemical Processes.* New York: AIChE.

CCPS. 1998a. *Guidelines for Improving Plant Reliability through Data Collection and Analysis.* New York: AIChE.

CCPS. 2000. *Guidelines for Chemical Process Quantitative Risk Analysis,* 2nd Edition. New York: AIChE.

CCPS. 2001. *Layer of Protection Analysis: Simplified Process Risk Assessment.* New York: AIChE.

CCPS. 2006. *Guidelines for Mechanical Integrity Systems.* New York: AIChE.

CCPS. 2007a. *Guidelines for Risk Based Process Safety.* New York: AIChE.

CCPS. 2007b. *Guidelines for Safe and Reliable Instrumented Protective Systems.* New York: AIChE.

CCPS. 2007c. *Human Factors Methods for Improving Performance in the Process Industries.* New York: AIChE.

CCPS. 2008a. *Guidelines for Hazard Evaluation Procedures,* 3rd Edition. Hoboken, NJ: John Wiley & Sons.

CCPS. 2008b. *Guidelines for the Management of Change for Process Safety.* Hoboken, NJ: John Wiley & Sons.

CCPS. 2009a. *Guidelines for Developing Quantitative Safety Risk Criteria,* 2nd Edition. New York: AIChE.

CCPS. 2009b. *Inherently Safer Chemical Processes: A Life Cycle Approach,* 2nd Edition. New York: AIChE.

CCPS. 2013. *Guidelines for Enabling Conditions and Conditional Modifiers in Layer of Protection Analysis.* New York: AIChE.

CCPS. 2013b. *Guidelines for Determining the Probability of Ignition of a Released Flammable Mass.* New York: AIChE.

IEC (International Electrotechnical Commission). 2003. *Functional Safety: Safety Instrumented Systems for the Process Industry Sector - Part 1: Framework, Definitions, System, Hardware and Software Requirements.* IEC 61511. Geneva: IEC.

IEC. 2010. *Functional of //programmable electronic safety related systems.* IEC 61508. Geneva: IEC.

SINTEF (Stiftelsen for industriell og teknisk forskning). 2010. *PDS Data Handbook.* Trondheim, Norway: SINTEF.

2

OVERVIEW: INITIATING EVENTS AND INDEPENDENT PROTECTION LAYERS

2.1   LOPA ELEMENTS: AN OVERVIEW

The reliability of both equipment and human performance is dependent upon the management systems that support their ongoing effectiveness. Inherent in the IEF and IPL PFD values provided in this document is an assumption that the equipment is well designed, installed, and maintained and that factors influencing human error are well controlled. Therefore, it is important to discuss some key elements of an effective process safety management program to ensure that use of the values provided in this document is appropriate for a specific facility.

Sound management systems form the foundation for LOPA to be effectively implemented in an organization. However, to appropriately use the IEF and IPL PFD values found in this document, it is also important to have an understanding of how these values will be applied within the context of the LOPA methodology. This chapter reviews some key aspects of the LOPA method and important elements to consider when selecting the appropriate scenarios, estimating each scenario's IEF, assessing the potential consequences of each scenario, and applying the appropriate number and type of IPLs to manage the risk.

2.2   MANAGEMENT SYSTEMS TO SUPPORT LOPA

LOPA is a risk assessment tool used to identify system features and actions that can lead to, or prevent, an incident. This allows an organization to focus on maintaining the reliability of the process equipment and human actions that prevent scenarios of concern. Identifying IEs and IPLs is important, but if the features of these systems are not maintained properly, the risk estimate may be in error and the scenarios of concern may occur more frequently than predicted.

16

To ensure that system elements perform as required, an organization should implement effective management systems that control, check, test, audit, review, and maintain their reliability. *Guidelines for Risk Based Process Safety* (CCPS 2007a) gives an excellent model for the overall structure of an effective management system program that is directly applicable to LOPA. Key elements of the management system program that affect IEs and IPLs address equipment reliability, performance tracking, performance management, work procedures, quality assurance, and change control. It is assumed that organizations using the IEF and IPL PFD values found in this document have already implemented effective process management systems. Otherwise, the IEF and IPL PFD values in this document may not be applicable to the specific site, and the actual level of process risk may be higher than estimated.

Effective management system programs include elements such as the following:

- *Procedures* – Written instructions can be used by operations, maintenance, and other support functions to ensure that tasks are completed properly. For procedures to be most effective, they are written in a manner that minimizes the probability of human error.

- *Training/Experience* – Unless workers are competent in their assigned responsibilities, the IEF and IPL PFD values applicable to that site may be significantly different from those listed in this document. Establishing effective initial and refresher training programs is important to ensure a high performing organization.

- *Auditing* – Auditing is used to ensure that procedures, training, maintenance, and other management systems are performing as expected. Ongoing performance measures can include key metrics that indicate the effectiveness of the organization's management systems.

- *Inspection, Testing, and Preventive Maintenance (ITPM)* – ITPM focuses on ensuring the integrity of the process equipment and the reliability of the instrumentation, controls, and other process IPLs. The IEF and IPL PFD values presented in this book are appropriate for an organization with an ITPM system in accordance with RAGAGEP.

- *Leadership and Culture* – Visible leadership is critical; it is central to the success of all management systems. Leadership commitment is crucial to drive a strong safety culture within an organization. One aspect of visible leadership is ensuring that sufficient resources are allocated to maintain IPLs at their expected level of performance. Actions such as engaging employees in safety activities, encouraging near-miss reporting, and acknowledging and rewarding safe behavior provide clear indication of leadership support for process safety.

- *Incident Investigation* – It is important to investigate incidents and share the lessons learned to prevent repeat events from occurring. *Guidelines for Investigating Chemical Process Incidents*, 2nd Edition (CCPS 2003) defines root causes in terms of gaps in management systems, and thorough investigations can result in the identification and correction of these gaps.

*   *Management of Change (MOC)* – The integrity of IEs and IPLs should be protected by a change management protocol. Safety performance can be quickly degraded if process changes, such as changes to materials, chemical composition, equipment, procedures, facilities, or organizations, occur without appropriate technical and risk reviews.

*   *Human Factors* – At a fundamental level, nearly all process safety incidents are caused by human error, since humans design, specify, receive, install, maintain, and operate process equipment. These errors may occur during any phase of the process lifecycle and can result from various management system failures. One goal of any management system is to minimize human error; this will reduce the frequency of IEs and help ensure the reliability of human IPLs.

*   *Impairment/Deficiency Procedures* – At some point, an IPL may be found to be in a deficient state. This may result in the need to temporarily bypass that IPL until repair or replacement is complete. Prompt action is important to minimize the time at risk while the IPL is impaired and to quickly restore the system to normal operation. When an IPL is known to be deficient, compensating measures equivalent to the lost risk reduction can be implemented until the IPL is restored. A root cause analysis of the deficiency can help to identify and correct management system gaps and prevent reoccurrence.

The importance of having strong process safety management systems cannot be overstated; they are essential for the effective implementation of LOPA. As stated in *CCPS LOPA* (2001):

> *"LOPA's cornerstone is the organization's policies and practices regarding risk management."*

2.3   SCENARIO SELECTION

Once effective management systems are in place and an organization has begun to use LOPA, an important initial step in application of the methodology is the identification of LOPA scenarios. A scenario is an event that could lead to an outcome of concern. In LOPA, each scenario is a unique cause-consequence pair, where one initiating event leads to a chain of events that can result in a specific consequence of concern. LOPA is not a technique for identifying scenarios. Rather, LOPA is a simplified method used to estimate the risk associated with a previously identified scenario and to ensure that there are sufficient IPLs in place to manage the risk. There are several means of identifying scenarios for analysis in LOPA:

- Structured hazard evaluation methods, such as Hazard and Operability Studies (HAZOPS)
- Semi-structured hazard evaluation methods, such as What-If Analysis or Checklist Analysis techniques
- Analysis of equipment or process failure modes with structured methods, such as Failure Mode and Effects Analysis (FMEA)
- Plant operational experience, which may identify potential processing deviations or failures not previously anticipated
- Review of plant and industry incident and near-miss data, which may reveal previously unidentified credible scenarios
- Management of change reviews, where new scenarios can be identified based on proposed changes to procedures, materials, processing conditions, equipment, or personnel

Inherently safer design principles can be used to remove scenarios from a LOPA by eliminating the potential for the event of concern to occur. This approach is discussed further in Chapter 4, and the reader is also referred to *Inherently Safer Chemical Processes: A Life Cycle Approach,* 2nd Edition (CCPS 2009b) for more guidance on how to apply inherently safer concepts to reduce process risk.

The scenarios of concern at a facility should be evaluated to ensure that there are adequate protection layers to prevent undesired consequences; however, analyzing more scenarios than necessary leads to an inefficient use of resources. One approach to reducing the number of LOPA scenarios is to analyze scenarios for one unit operation (usually the worst case) and then apply the same IPLs to similar unit operations. For example, scenarios associated with one storage tank can be assessed, and the protection layers identified can be used to protect other lower risk storage tanks. Similarly, some companies will analyze one scenario resulting from a sample valve being left open to represent all sample valves on the same system. This approach leads to efficiency of analysis and consistency of IPLs used. It can also be overly conservative and result in more IPLs applied than required for unit operations with lesser hazards.

Once a scenario is chosen for analysis in LOPA, IPLs are identified that will reduce the risk of the event progressing to the consequence of concern. However, to properly select IPLs, it is necessary to understand how each IPL is intended to function in the scenario. Some IPLs are used to prevent the consequence from occurring; these are known as preventive IPLs. An example of a preventive IPL is a high level switch that stops flow to the tank and prevents an overflow. Other IPLs are used to reduce the severity of the consequence; these are known as mitigative IPLs. Examples of mitigative IPLs include a dike that minimizes environmental impact or an excess flow check valve that reduces the quantity of material released to the surroundings upon loss of containment.

Preventive and mitigative IPLs may effectively prevent the specific consequence of concern that is being analyzed in a particular scenario. However, successful use of a preventive IPL may also generate a different, but still significant, consequence. For example, successful activation of a relief device may prevent vessel overpressure, but the activation may create a new release scenario when the material is vented to another location. Similarly, for mitigative IPLs, a dike may minimize an environmental impact due to a liquid spill; however, other scenarios may be created, such as a toxic cloud evaporating from the dike contents or a pool fire if spilled flammable material ignites.

When IPLs are used in LOPA, consideration should be given to the scenarios that could occur if the IPLs function as intended. These scenarios may also need to be evaluated using LOPA to ensure that all residual risks are adequately managed.

2.4   OVERVIEW OF SCENARIO FREQUENCY

Once a list of scenarios has been created, the IEF values for each scenario can be developed. The most common initiating events for LOPA scenarios relate to equipment failure and human error. This section discusses factors influencing equipment failure rates and the probability of human error. Also discussed are sources of data that can be used to develop IEFs and IPL PFDs as well as to validate those values that have been applied in previous LOPAs.

2.4.1   Scenario Definition and Level of Analysis

Scenarios of concern are identified using hazard evaluation processes. LOPA is used to evaluate these scenarios by examining how initiating events propagate to an undesired consequence. IEs can be defined at any level of detail. The analysis level needed is generally limited to what is necessary to understand the required effectiveness of the IPLs and the level of independence that exists between the IE and IPLs. The various equipment and human causes of an IE are often discussed to understand the underlying system causes for incident occurrence and to assist in the validation of IEF and IPL PFD estimates. While each cause could be developed into a separate scenario, the risk criteria are often defined for a loss event, or for the harm posed by the loss event. The method for estimating the scenario frequency should be consistent with the organization's risk criteria.

For example, consider a loss event initiated by failure of a pressure control loop that allows the process pressure to exceed the vessel maximum allowable working pressure (MAWP). The pressure control loop malfunctions due to a failure of the control system. The specific causes of the control system failure may be

- Failure of its components, such as field sensor, controller, or final element
- Failure of utilities, such as power supply or instrument air
- Failure of people, such as operators or maintenance staff, to follow procedures
- Failure of support systems, such as purges or heat tracing
- Failure of interfaces, such as the human machine interface, communications, or security provisions
- Errors of commission and omission, such as design, configuration, or management of change errors

Examining individual causes ensures that the appropriate IEF values are used and that effective IPLs are chosen. However, as illustrated above, the number of specific causes can be quite numerous, and analysis of these detailed scenarios may not be value-added. To conduct a LOPA efficiently, it is important to be selective and to perform individual cause-consequence analysis on those scenarios that truly add value to the risk assessment.

2.4.2   Equipment Failure Rate Considerations

Managing equipment performance is important to prevent equipment failure from initiating LOPA scenarios and from adversely affecting IPL PFDs. The reliability of equipment over time is sometimes portrayed using a "bathtub" curve, as shown in Figure 2.1. The bathtub curve illustrates three regions in the lifecycle of equipment.

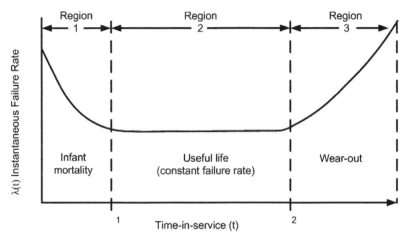

Figure 2.1. Bathtub curve (CCPS 2000).

2.4.2.1 Early-Life Period

New components are often introduced to a process during construction, upgrade, and replacement. Immediately after installation, component failure rates can be higher due to manufacturing errors, incorrect material of construction, damage during shipment, and installation and commissioning errors. This phase of early failures, represented in Region 1, is sometimes referred to as "infant mortality" or "burn-in." Inspection, commissioning, and validation activities are used to identify and correct these failures.

2.4.2.2 Constant-Rate Period

Generally, the early-life period is relatively short. This is followed by an extended time (commonly considered the useful life of the component), during which the average failure rate is much lower than in early life and is relatively constant. Failures during this "constant-rate" phase, represented by Region 2, tend to be more random in nature.

2.4.2.3 Wear-Out Period

As components age, failure rates begin to rise as the unit approaches its end of life. This period of increasing failure rate over time, represented by Region 3, is sometimes referred to as the "wear-out period."

> For the purpose of LOPA, when assigning a failure rate, the assumption is made that the device is in the constant-rate period, where failures are random.

For equipment failure data, it is generally assumed that there is an effective asset integrity program in place to ensure that the equipment remains within its useful life. The result is that, following a proof test, the equipment is returned to its *like new* or *as good as new* condition. The implication of *like new* is that the equipment continues its useful life (on the flat part of the bathtub curve, where infant mortality is not an issue) and will not reach wear-out prior to its next proof test.

2.4.3   Human Error Rate Considerations

Managing human performance is important to prevent errors that can initiate LOPA scenarios and adversely impact the reliability of IPLs. Human error depends on a number of factors that should be considered during the selection of IEF and IPL PFD values. These factors include

- *Procedure accuracy and procedure clarity* – Do procedures have a high level of accuracy, do they clearly convey the information, and are they convenient to use?

- *Training, knowledge, and skills* – Has selection criteria for new employees been established that ensure that personnel are capable of becoming qualified for the duties they are to perform? Effective, demonstration-based initial and refresher training can be used to develop and maintain skill level.

- *Fitness for duty* – Have factors such as fatigue, stress, illness, and substance abuse been managed during all phases of operation? It is important that workers be physically capable of completing the tasks required.

- *Workload management* – Has workload been optimized during all phases of operation, including normal, startup, and emergency shutdown modes? If the workload is too low, operators may become bored, resulting in decreased vigilance. If an operator's workload is too high, however, human error will tend to increase as a result of task overload.

- *Communication* – Have management systems and protocols for proper communication on radios and during shift changes been implemented? Miscommunication is a frequent cause of human error in the workplace, and effective communication strategies can reduce human error due to miscommunication.

- *Work environment* – Have factors such as lighting, noise, temperature, humidity, ventilation, and distractions been managed to minimize their contribution to human error?

- *Human-machine interface* – Does the human-machine interface (HMI) facilitate the operator's interaction with the process? The layout of equipment, displays, and controls strongly affects human performance. Abnormal conditions need to be clearly annunciated, and alarm management is important to prevent nuisance alarms or alarm overload.

- *Job complexity* – Has the job been assessed to ensure that the task is not overly complex, making human error more likely to occur? The complexity of an activity is proportional to several factors, including the number of steps, the level of difficulty of the procedure, and the degree to which judgment or calculation is required. More information on the science of human factors and the influence of performance shaping factors on human error rates can be found in *Guidelines for Preventing Human Error in Process Safety* (CCPS 1994). Also refer to Appendix A of this document for additional discussion on human factors.

2.4.4   Failure and Error Rate Data Sources

The majority of equipment failure data that exist are component data. Data range from "soft" to "hard." "Soft" data originate with "expert judgment" or generic experience. More specific data come from plant experience, with varying degrees of confidence in values, depending on the nature of the information available. At the other end of the spectrum, "hard" data consist of systematic collections and analyses of plant event data, component failure data, and detailed studies of human reliability.

Equipment reliability data and human reliability data may be obtained from many sources. The quality and type of each source is sufficiently different such that significant conflicts are likely to exist among the different sources. To arbitrarily select published values would result in the misuse of the data that exist. In order to use any data source, the analyst should understand the fundamental basis of the data in order to determine applicability to the specific process under evaluation. This section is intended to help the reader understand the various types of data so that an appropriate determination of the dataset's applicability to the study can be made.

2.4.4.1 Differences in Data

Differences in data exist for a number of reasons and result in a range of values. For equipment failure data, these reasons include, but are not necessarily limited to

- Process service and application (e.g., corrosive, fouling)
- Installed environment (e.g., ambient temperature, exposure to the elements, chemical exposure)
- Operating mode (e.g., continuous operation, cyclic operation, standby)
- Failure mode of interest (Some data collection systems and generic data sources do not adequately distinguish failure modes.)
- Maintenance strategy (e.g., run to failure, preventive maintenance, condition-based maintenance)
- Site population (Is the tracked dataset large enough to be statistically valid for calculating failure rates?)
- Quality of the event data (Is the data sufficient to support the IEF and IPL PFD claims?)

Similarly, for human reliability data, the reasons for differences in data include

- Quality of procedures (Are the procedures accurate and easy to understand?)
- Effectiveness of training (Has the worker been adequately trained on the procedure to perform a specific task?)
- Human factors engineering (e.g., labeling, display design, alarm management, and valve placement)
- Work situation and environment (e.g., control of distractions, including temperature, lighting, and noise)
- Fitness for duty (e.g., controls against fatigue, stress, health issues)

- Controls against miscommunication (e.g., radio use protocol, shift change records)
- Measurement of human error rates (e.g., documenting the number of times that a task has been done as well as the number of errors that have occurred)

2.4.4.2 Data Sources

Data sources can be categorized as

- Expert judgment
- Generic
- Predicted
- Site-specific

The following sections describe each category of data source:

- *Expert Judgment* – Faced with limited resources and insufficient data collection and data mining capabilities, analysts sometimes make use of expert judgment. Expert judgment may also be a viable alternative for getting an order of magnitude estimate of equipment failure rates and human error rates. Expert judgment entails consulting with acknowledged experts in the field of interest. The experts can provide estimates of equipment reliability or human reliability based upon their personal or collective experiences. Expert judgment can be derived via a group of experts in more than one way, and some approaches can help to eliminate bias or memory loss that may otherwise be present. One approach is the Delphi Method (Linstone 1975). A more recent approach is provided in *Eliciting and Analyzing Expert Judgment* (Meyer and Booker 2001). Expert judgment can result in the selection of values better than those derived from generic or predicted data. The best use of expert judgment may be to offer an opinion on the validity of equipment reliability data or on human reliability data obtained via other means. The judgment of experts can help in the appropriate selection of values from the other data sources and determination of the quality of that data. Some of the values in this text have been derived from expert judgment and consensus among the *Guidelines* subcommittee when values were not available from one of the other data sources. These values are supported by the evaluation of the many experts present on the *Guidelines* subcommittee.
- *Generic* – Typically publicly available, generic data have been aggregated from similar systems or situations. Often, this aggregation tends to lessen the usefulness of the data. Detail can be lost in the process of merging datasets together or converting raw data to values included in simplified tables. Generic data are available from a variety of sources. Chapter 4 of *Guidelines for Process Equipment Reliability Data, with Data Tables*

(CCPS 1989) contains numerous descriptions and references to generic equipment failure data sources. The various phases of data within the Offshore Reliability Data 1984-2009 (OREDA 2009) are excellent sources of failure rate values for oil and gas (and other petrochemical) facilities. U.S. Nuclear Regulatory Commission (NRC) documents, including NUREG CR-1278 (Swain and Guttmann 1983), contain generic data, expert judgment data, and predicted data for human error rates as originally developed from U.S. military and nuclear power plant sources.

Sources of generic data include vendors of components. Such data are not always generated from bench test data; they often come from return/repair data. Customers might not report failures of lower value equipment to the vendor. As a result, failure data collected by the vendor for some components might be overly optimistic. The data may not reflect how the component will perform in a specific operating environment (e.g., a corrosive atmosphere). Vendor data also commonly exclude failures that were not the "fault" of the component (e.g., dirty instrument air, voltage spikes). Therefore, it is important to understand the basis for vendor data as well as the process in which the component will be used.

- *Predicted* – Prediction refers to the application of basic failure rate data for the elemental components to determine the failure rate of the aggregate system or the error rate for a given task. Elemental equipment components include items such as contacts, coils, springs, capacitors, transistors, resistors, bearings, etc. High-quality data are more likely to exist for such elemental components due to the need for manufacturers to perform lifecycle tests on their products. Data provided by the manufacturers are only applicable to the system included within the manufacturer's boundary diagram. The data do not account for specific process/environmental conditions or site-specific work practices that may affect the actual failure rates of the system.

  Human reliability can be predicted using a number of techniques, such as Standardized Plant Analysis Risk-Human Reliability Analysis method (SPAR-H) (Gertman 2005), Human Event Repository and Analysis (HERA) (Hallbert et al. 2007), Technique for Human Error Rate Prediction (THERP) (Swain and Guttmann 1983), and Human Error Assessment and Reduction Technique (HEART) (Kirwan 1994). These methods have been developed from various data sources and assess the probability of human error based on the impact of various performance shaping factors. For more information on performance shaping factors, refer to Appendix A of this document.

- *Site-Specific* – The ideal data for an analysis is specific to the plant and to the application being analyzed. It is also extensive enough for a statistically significant analysis. For equipment failure rates data, installations are in service long enough to ensure adequate operational history. Similarly, for human error data, estimates are based on multiple operators monitored for a sufficient period of time to yield confidence that the data are relevant to the average operator or maintenance person. Site-specific data demonstrating reliability is often referred to as "proven in use" or "prior use."

Equipment failures can be revealed or unrevealed. Revealed failures are clearly evident through process deviation, alarms, or other indication. Revealed failures can be corrected in a reasonably short period of time. Unrevealed failures are not immediately evident; they are dormant until discovered, usually by diagnostic testing or routine maintenance. For equipment with revealed failures, data may include a clear description of the equipment, the number of operating years of experience, and the documentation of event data with sufficient detail to determine if specific failure modes have occurred. The population of data is of similar design and application, and it is subject to the same level of maintenance as the systems being studied. For equipment providing protection where failures are unrevealed, it is also important to consider the proof test frequency for each protection item, since the statistical techniques for evaluating revealed failure populations versus unrevealed populations are quite different. Refer to Appendix C of this document for additional information.

Readers wanting to know more about developing high-quality, reliable data for equipment and measuring performance on a cost-effective basis are referred to *Guidelines for Improving Plant Reliability through Data Collection and Analysis* (CCPS 1998a) as well as to the website describing the CCPS's ongoing Process Equipment Reliability Database (PERD) initiative:

http://www.aiche.org/CCPS/ActiveProjects/PERD/index.aspx

Similarly, for human reliability data, the best data are collected within the specific unit. This can be accomplished by recording and analyzing the data for the number of errors recorded versus the number of executions of a task. Tests or drills are also an effective way to measure many human response error rates; more details are given on these approaches in Appendix B of this document.

When developing the data for use by a particular company, and sometimes for a particular site within that company, it is generally necessary to leverage all of the above categories of data sources. Companies are encouraged to consolidate process performance data to provide a better source of site-specific failure rate data.

2.4.5   Validation of Failure/Error Rate Data

Validation is an assessment to confirm that the IEF or IPL PFD value used in the LOPA is appropriate for a given scenario. As plants develop operating history, or as process changes occur, it may be necessary to validate that the values previously chosen for use in LOPA are still applicable. The claimed IEF or IPL PFD value could be confirmed using one of the methods discussed in the previous section.

Some IEF and IPL PFD values presented in the data tables of this document are based on published data or the expert opinion of the *Guidelines* subcommittee members. When these values are used in LOPA, it is assumed that the design, operation, and maintenance practices associated with the IEF or IPL PFD values are aligned with the guidance provided with the generic values. Verification that the site is following the guidance associated with the suggested values in the data tables is important to ensure that the values selected are valid for the given scenario.

2.5   OVERVIEW OF CONSEQUENCES

Risk is a function of both the frequency of occurrence of an incident scenario and the potential impact of the resulting consequence. Therefore, assessing the appropriate consequence severity is as important as assessing the frequency of the scenario to correctly estimate the level of risk.

Below is a brief discussion of the evaluation of consequence severity. Detailed coverage of consequence estimation is beyond the scope of this document. Guidance was provided in *CCPS LOPA* (2001), and detailed consequence evaluation methods are presented in *Guidelines for Consequence Analysis of Chemical Releases* (CCPS 1999).

Just as the principles of inherently safer design can be used to reduce the frequency of LOPA scenarios or eliminate them entirely, the application of good design practices can also be used to reduce the consequences of scenarios of concern. This may reduce the number of IPLs that are needed to mitigate a specific event scenario.

2.5.1   Evaluation of Consequence Severity

In LOPA, a consequence of concern is the ultimate outcome of a LOPA scenario that is unmitigated by any IPLs. Consequences of concern to an organization can include toxic exposure, fire, environmental impact, business loss, and other outcomes. The worst credible consequence of an event is generally assessed, and scenarios are selected for LOPA based on the individual organization's protocol for selecting scenarios.

There are two basic approaches to estimating the consequence severity:

- Classify the release into a *consequence category*, based on factors such as the estimated amount of material released and its chemical and physical properties. Examples can be found in Tables 3.1 and 3.2 in *CCPS LOPA* (2001).
- Estimate the *severity* resulting from the scenario of concern, usually in terms of impact, such as the numbers of fatalities, level of environmental

impact, or financial loss due to equipment damage or loss of production. Assessments of potential impact severity can be performed by modeling potential consequences. Depending on the type of impact of concern to the organization, the use of conditional modifiers may also be used for estimating the scenario frequency. Refer to *Enabling Events and Conditional Modifiers in Layer of Protection Analysis* (CCPS 2013) for additional information on this topic.

In the *CCPS LOPA* (2001), these approaches were illustrated using four consequence estimation methods. Regardless of the approach, a standardized methodology is generally adopted within an organization to promote consistency of application.

2.5.2 Inherently Safer Design and Consequence Severity

As discussed earlier in this chapter, inherently safer design practices can be used to eliminate scenarios or reduce the potential consequence severity. Examples of process modifications that can reduce consequence severity include the following:

- Minimizing the chemical inventory in a process by decreasing line and equipment sizes or otherwise reducing the quantity of material in the process equipment.

- Moderating the conditions of the process, such as by operating at reduced temperatures and pressures or operating a more dilute system, to reduce the potential consequence severity.

- Substituting a less hazardous material can reduce the consequence of a release. Examples of this include using diluted materials (e.g., using 28% ammonia instead of anhydrous ammonia) and substituting a less hazardous chemical (e.g., using a hypochlorite solution instead of chlorine gas).

- Limiting the quantity of a reagent in a reactor by gradually feeding the material, rather than adding it in one charge, can reduce the potential consequence associated with an uncontrolled reaction.

- Reducing the impact of a fire or explosion on multiple receptors by the use of proper equipment spacing and facility siting. (This is also referred to as reducing "knock-on" effects.)

2.6   RISK CONSIDERATIONS

As previously discussed, LOPA is a simplified technique for assessing risk that is applicable in many situations. There are times, however, when other risk assessment methodologies can be used to supplement LOPA or can be used instead of LOPA:

1.   LOPA is not the right risk assessment methodology – LOPA can be used effectively to assess risk in many applications. However, LOPA may not be the best tool to assess the risk of some scenarios, such as those involving natural disasters or acts of terrorism.
2.   There is lack of independence – The LOPA methodology gives credit only for safeguards that are independent from each other. However, other approaches can estimate the level of risk reduction that can be achieved using a combination of safeguards that are not completely independent.
3.   Validation of IEFs or IPL PFDs is required – Methods such as FTA can be used to assess the failure rates of various components of an IPL and to validate that the system meets the PFD assumed in LOPA. Likewise, HRA methods can be used to evaluate IPL PFDs involving human response.
4.   There is a need for more rigorous analysis – LOPA is generally practiced as a simplified order of magnitude technique. As such, LOPA can yield conservative results. Thus, when LOPA indicates that the organizational risk criteria have not been met, other quantitative methods may be used to determine whether additional risk reduction measures are needed.

2.6.1   Risk Assessment Methodologies

There is a range in the complexity of methods available to quantify the risk of incident scenarios, from LOPA to Chemical Process Quantitative Risk Assessment (CPQRA). The term "quantitative risk assessment" (QRA) actually refers to a host of methods that may be used in concert with each other. The methods may generate a more detailed calculation of an IEF or IPL PFD. These approaches may also evaluate a range of consequences, such as the impacts to individuals in the assessment of individual risk and the impact to groups of people when evaluating societal risk. Generally, other QRA methods are more complex than LOPA, and they allow use of factors and safeguards that are typically not used in LOPA. A discussion of the application of QRA methods can be found in Chapter 6.

2.6.2   Risk Criteria

Risk management at an operating site is the product of many individual decisions. Some decisions can be made in a qualitative fashion, but others may warrant evaluation against more quantitative criteria. Like all risk estimation methods, the use of LOPA requires defined risk criteria to determine if the risk is tolerable to an organization. Risk criteria normally include a numerical target for maximum level of risk, or a risk matrix displaying how much risk reduction is required for a particular frequency/consequence severity ranking. *CCPS LOPA* (2001) gives examples of risk criteria used by several companies and several regulatory bodies. *Guidelines for Developing Quantitative Safety Risk Criteria* (CCPS 2009a) provides more rigorous treatment of risk criteria and its meaning relative to individual risk and societal risk.

It should be noted that the tolerable risk criteria selected by organizations can differ. Some organizations select different endpoints for analysis, which can vary from loss of containment to one or more fatalities. Some organizations use enabling conditions, conditional modifiers, or actual (rather than rounded, order of magnitude) values in LOPA; others do not. Regardless of the approach, it is important to ensure that the risk assessment methodology used and the tolerable risk criteria against which the results are measured are aligned and applied consistently.

2.7 CONCLUSIONS

Strong management systems are the foundation of LOPA. To support the values used for IEs and IPL PFDs in LOPA, effective management systems should be implemented and maintained to ensure the ongoing reliability of equipment and human performance.

LOPA is not a scenario identification method; when an organization chooses to use LOPA, the appropriate cause-consequence pairs are generally identified using hazard identification techniques. The IEF and IPL PFD values selected for each LOPA scenario can be obtained using expert judgment, generic data, predictive methods, or data generated for a specific site. These same data sources can also be used to validate IEF and IPL PFD values that have been used in previous LOPA analyses. An assessment of the scenario consequence will give an indication of the potential severity of an event, and the reader is urged to consider the use of inherently safer principles to reduce the potential consequences that could result from an incident scenario.

Ultimately, risk is a function of both the scenario frequency and the potential consequence severity. The risk determined by LOPA is typically compared against a corporate risk target or set of risk tolerance criteria to determine the number of IPLs needed to meet a risk goal. Every risk mitigation strategy has benefits and residual risks. The goal is to select the optimal solution while minimizing and managing the residual risks that remain. LOPA studies can be used to assess these residual risks to ensure that they are adequately controlled.

The remaining chapters will address the primary objective of this document, which is to provide additional guidance related to the selection of appropriate IEF and IPL PFD values for LOPA.

REFERENCES

CCPS. 1989. *Guidelines for Process Equipment Reliability Data, with Data Tables.* New York: AIChE.
CCPS. 1994. *Guidelines for Preventing Human Error in Process Safety.* New York: AIChE.

CCPS. 1998a. *Guidelines for Improving Plant Reliability through Data Collection and Analysis.* New York: AIChE.

CCPS. 1999. *Guidelines for Consequence Analysis of Chemical Releases.* New York: AIChE.

CCPS. 2000. *Guidelines for Chemical Process Quantitative Risk Analysis,* 2nd Edition. New York: AIChE.

CCPS. 2001. *Layer of Protection Analysis: Simplified Process Risk Assessment.* New York: AIChE.

CCPS. 2003. *Guidelines for Investigating Chemical Process Incidents,* 2n^d Edition. New York: AIChE.

CCPS. 2007a. *Guidelines for Risk Based Process Safety.* New York: AIChE.

CCPS. 2009a. *Guidelines for Developing Quantitative Safety Risk Criteria,* 2nd Edition. New York: AIChE.

CCPS. 2009b. *Inherently Safer Chemical Processes: A Life Cycle Approach,* 2nd Edition. New York: AIChE.

CCPS. 2013. *Guidelines for Enabling Conditions and Conditional Modifiers in Layer of Protection Analysis.* New York: AIChE.

CCPS. n.d. Process Equipment Reliability Database (PERD). http://www.aiche.org/ccps/resources/perd.

Gertman, D., H. Blackman, J. Marble, J. Byers, and C. Smith. 2005. *The SPAR-H Human Reliability Analysis Method.* NUREG CR-6883. Washington, DC: U.S. Nuclear Regulatory Commission, Office of Nuclear Regulatory Research.

Hallbert, B., A. Whaley, R. Boring, P. McCabe, and Y. Chang. 2007. *Human Event Repository and Analysis (HERA): The HERA Coding Manual and Quality Assurance* (NUREG CR-6903, Volume 2). Washington, DC: Division of Risk Assessment and Special Projects, Office of Nuclear Regulatory Research, U.S. Nuclear Regulatory Commission.

Kirwan, B. 1994. *A Guide to Practical Human Reliability Assessment.* Boca Raton, FL: CRC Press.

Linstone, H. 1975. *The Dephi Method.* Boston: Addison-Wesley.

Meyer, M., and J. Booker. 2001. *Eliciting and Analyzing Expert Judgment.* Philadelphia: Society for Industrial and Applied Mathematics.

OREDA. 2009. *Offshore Reliability Data Handbook,* 5th Edition *(OREDA).* Trondheim: SINTEF.

Swain, A., and H. Guttmann. 1983. NUREG CR-1278 *Handbook of Human Reliability Analysis with Emphasis on Nuclear Power Plant Applications.* Washington, DC: Sandia National Laboratories.

3

CORE ATTRIBUTES

3.1 INTRODUCTION TO CORE ATTRIBUTES

Guidelines for Safe and Reliable Instrumented Protective Systems (CCPS 2007) outlined seven core attributes that are basic characteristics of an effective IPL. The principles associated with these core attributes can also influence the frequencies of initiating events. The values given in the data tables in this book presume that these core attributes are incorporated into the design, operation, and maintenance of the systems being evaluated. If these fundamental attributes are not present, the values presented in this document may not be valid. Therefore, it is appropriate to review certain key aspects of these attributes to ensure that the IEF and IPL PFD values in this document are appropriately applied to a site-specific LOPA.

The seven core attributes are

1. Independence
2. Functionality
3. Integrity
4. Reliability
5. Auditability
6. Access security
7. Management of change

An overview of these core attributes can be found in *Guidelines for Safe and Reliable Instrumented Protective Systems* (CCPS 2007) and will not be repeated in this document. However, certain critical aspects of these attributes are discussed in this chapter to help the reader determine whether the IEF and IPL PFD values presented in this document are applicable to a specific scenario.

The core attributes focus on characteristics that are actively managed by the site to achieve the required performance. These attributes will be discussed individually below, with an emphasis on the design features and management practices that affect each attribute. There is also quite a bit of overlap between these attributes. The strength, as well as the weakness, of any attribute affects the achievable performance of the overall system, including aspects of access security impact reliability, integrity, management of change, and auditability.

One type of human error that can affect multiple core attribute is referred to as *systematic* (also called *systemic*) error. Systematic errors do not occur randomly; rather, they result when the human error occurs. Systematic errors persist until identified and corrected. ISA-TR84.00.02 (2002) defines systematic error as an error that occurred during specification, design, manufacturing, implementation, installation, commissioning, operation, or maintenance. Systematic errors can affect both initiating events and IPLs; for example, poor maintenance can impact both IEF and IPL PFD values.

The probability of systematic error is reduced by implementing rigorous management systems, such as those discussed in Chapter 2, and by ensuring that the fundamental core attributes listed in this chapter are in place and remain effective. Examples of the types of systematic failures to consider in developing an IPL management system include

- Initial failure at startup (incorrect specification, incorrect calibration, improper material received, improper installation, wrong set point, etc.)
- Safety instrumented systems being left bypassed
- Pressure relief valves being left isolated from the process
- Delays in repair following failure detection
- Faulty preventive maintenance or repair
- Ineffective management of change

For a more complete discussion of the nature of equipment failures, the reader is referred to *Guidelines for Improving Plant Reliability through Data Collection and Analysis* (CCPS 1998a), especially Chapters 1 through 3.

3.2   INDEPENDENCE

Independence is a basic tenet of LOPA. Independence is achieved when the performance of one IPL is not affected by the IE or by the failure of any other IPL to operate. When the effective performance of one component is *dependent* upon the successful operation of another, the devices are not independent.

Independence is an important concept, although absolute independence is generally not achievable. Plants have shared utilities, generally one maintenance staff, common calibration devices, and standard vendors who supply a number of similar components for applications throughout the facility. However, IPLs should be sufficiently independent such that the degree of interdependence is not statistically significant.

3.2.1   Dependent Safety Systems

IPLs are independent when their ability to prevent a consequence of concern is not impacted by the IE or failure other IPLs. There are situations, however, where a safety system can successfully act only if another system is at least partially effective. Some common examples include

- A pressure relief device on an upstream low pressure system is not sized for full reverse flow from the high pressure system. Instead, it is sized to handle flow based on 10% of the piping diameter, assuming that an upstream check valve will at least partially function to prevent backflow. (For additional discussion on this topic, refer to Section 5.2.2.2.2.)

- A pressure relief device is sized for the fire case assuming that cladding is in place to limit the heat flux into the vessel.

In these cases, the safeguards described do not represent independent IPLs. The successful prevention of the consequence of concern is dependent upon both components functioning. Instead, both components should be considered as one IPL, and the PFD claimed can be no less than the PFD of the least reliable component in the system.

3.2.2   Common Cause Failure

Some failures initiate multiple LOPA scenarios within a process. Other events can result in multiple, simultaneous failures of what appear to be independent pieces of equipment; this is one type of *common cause failure*. A common cause exists when one event results in the failure of more than one device, procedure, or system. (*Common mode failure* is a specific type of common cause failure, where one event results in the failure of more than one device, procedure, or system, and the devices fail in the same manner.) Neglecting to account for common cause failures can lead to incorrect LOPA results. Thus, the simplest LOPA approach is to ensure independence between the IE and an IPL (or between multiple IPLs) in the same LOPA scenario.

Typical rules for independence of an IPL from the IE and from other IPLs in the same LOPA scenario include

- The device or system implementing each IPL is independent of the failure of the equipment that causes the IE and of the failure of other IPLs credited in the scenario.

    The boundary for IEs and IPLs includes all devices, process connections, human interfaces, and support systems that operate when required to stop the event propagation. A lack of independence can occur when multiple instruments or devices are attached to the process via shared piping or tubing. The potential impact of these shared connections to the achievable performance needs to be carefully assessed. For example, a pressure relief valve (PRV) and a pressure switch sharing the same nozzle may be susceptible to some of the same failure modes, such as the plugging or closing of a shared valve.

Where humans interact with the process, the design and management of alarm interfaces, bypass capabilities, isolation valves, and manual shutdown systems should be assessed to understand how human factors, access security, and management of change are addressed. When the same personnel, procedures, and testing schedules are used to maintain multiple instruments, there is a greater potential for human error to adversely impact multiple layers of protection. To address these issues, it is important that effective process safety management systems are in place to ensure that the equipment is well designed, installed, and maintained and that factors influencing human error are well controlled.

Examples where the risk of common cause failure may be higher than expected:

- o Redundant instruments are serviced with the same person performing the calibrations (e.g., a person could miscalibrate both instruments) or using the same calibration instruments. (Faulty calibration instruments can cause both field instruments to be miscalibrated.)
- o The same type of valve, sensor, or other device is used in multiple systems. (In this case, a specification error or functional deficiency in the supplied devices can cause simultaneous failures of different systems.)

• The utility or support system implementing each IPL is independent of the failure of the equipment that causes the IE as well as of other IPLs.

A failure of a utility or support system that causes the IE should not be required to be operational for a claimed IPL to function. Utility failures can be a source of common cause failure; power outages can be initiating events and can also impair multiple safeguards that might otherwise be considered for use as IPLs. These types of events can induce the simultaneous failure of more than one instrumented system. Similarly, the failure of a support system that renders one IPL inoperative should not also disable another IPL protecting against the same scenario.

Example: When the operation of the IPL requires air, loss of air should not be the initiating event of the LOPA scenario or affect another IPL in the same scenario.

- A human (or group of humans that routinely work together) is used only once in a scenario (either as part of the IE or as part of an IPL).

In situations where it is desired to take credit for the action of multiple layers relying on humans, it is important to determine if the humans are relying on independent information and are expected to take action independently. Generally, workers from the same work group, or who are working as a team, are not considered to be independent of each other. If the humans are not fully independent, then use of a human reliability analysis (HRA) or other similar methodology may be needed to substantiate the credit being claimed.

Example: When operator error is the initiating event, it is generally assumed that the same operator does not self-correct or act to stop the scenario. Human error is comparable to the failure of a system. The human error may have been due to factors, such as inexperience, faulty operating instructions, distraction, work overload, fatigue, or other incapacity, that could also be present later when additional action is required of the operator. Once a human has committed an error, it is not reasonable to expect the same operator to act correctly later, as part of the same activity, without an independent trigger. When considering operator IPLs, the operating procedure and the information or data that the operator will rely upon to trigger action should be assessed to ensure that they are sufficiently independent of the IE to support the IPL claim.

In addition to equipment failure and human error, common cause failures can also occur due to natural events, such as fires, floods, hurricanes, tornadoes, earthquakes, or lightning strikes. Therefore, it is important to consider whether a common cause failure could both initiate a scenario and impair an IPL or whether common cause failure could impair the performance of multiple IPLs in a scenario.

3.2.3   Common Cause Modeled as Initiating Events

In some cases, LOPA can be used to address common cause issues by creating new scenarios using the common components as initiating events. For example, a control loop could fail due to a control valve malfunction, or it could fail due to a controller failure that affects multiple control functions simultaneously within the control system. To assess this common failure within a single scenario, separate LOPA scenarios could be created for control valve malfunction and for controller failure. In the controller failure scenario, IPL PFD credit would not be given for other control functions that would fail at the same time as the initiating event.

The overall failure rate of an initiating event is composed of the failure rates of the individual components and the failure rate associated with systematic error.

The generic IEF values proposed in this document incorporate common cause failure considerations and are generally more conservative than the sum of the failure frequencies of the discrete components. IEC 61511 (2003) limits risk reduction claims of SCAI implemented within a single BPCS to one order of magnitude, in part to account for systematic error. When creating separate LOPA scenarios to assess individual components of equipment that can initiate an event of concern, such as a control loop, it is important to ensure that the systematic failure rate is included in the LOPA analysis or the IEF value may be overly optimistic.

Some companies have geographic or personal risk criteria that require that the scenario frequencies from multiple LOPA scenarios be combined to calculate the overall risk. While breaking down initiating events into individual component failures can understate IE failure rates, aggregating scenario frequencies can result in overstating IE failure rates. This can occur when common cause failure is accounted for in multiple scenarios that are then combined. For example, there may be a number of scenarios that are initiated by the failure of control loops implemented in BPCS. The IEF of each of these scenarios includes a contribution for the failure rate associated with the BPCS logic solver. If multiple loop failure scenario frequencies are combined, the failure rate contribution of the logic solver to each scenario is aggregated. The resulting IEF total can overstate the contribution of the logic solver and can lead to the IEF erroneously exceeding the overall control system failure rate. Therefore, when aggregating LOPA scenario IEF, it is important to consider whether failures associated with common components are being over-counted.

3.2.4 Advanced Methods for Addressing Common Cause

Detailed analysis of common cause failures may be considered for the IE and IPLs where there is not complete independence. In these cases, the LOPA team may wish to turn to quantitative risk assessment techniques such as fault tree analysis (FTA), event tree analysis (ETA), or human reliability analysis (HRA) for supplemental analysis. In these techniques, common cause failure can be approximated mathematically. A *beta factor* is a factor used in FTA to account for the fraction of the total failure rate that is due to dependent or common cause failure. The beta factor is applied to calculations involving the simultaneous failure of two or more components whose failures are not statistically independent. This has the effect of increasing the calculated failure rate (or IPL PFD) of the system. Depending on the degree of dependency between components, the value for the IE failure rate (or IPL PFD) of the overall system can be dominated by common cause failure. Refer to *Guidelines for Chemical Process Quantitative Risk Analysis* (CPQRA), 2nd Edition (CCPS 2000) and Chapter 6 for additional guidance.

3.2.5   Common Cause Reflected in the Data Tables

Some common cause considerations are embedded in the IE and IPL guidance provided in this document. The limitations on various IEs and IPLs are discussed throughout Chapters 4 and 5. In some cases, the impact of common cause failure increases the PFD that could be claimed for a system that may have some level of dependency. A specific example is the suggested IPL PFD values for dual pressure relief valves in series. Although both pressure relief valves may be capable of independently protecting the process vessel and providing 100% of the required relief capacity, they may be of the same design, supplied by the same manufacturer, installed in the same process environment, and inspected and tested at the same time by the same staff. In addition to these potential common cause failure modes, there may be a practical, systematic limit to the PFD that can be claimed for two IPLs of a common technology. This is particularly true at low PFD values, where the beta factor can dominate the calculation and restrict the overall claim. Therefore, the level of protection provided by the two valves is less than what would be predicted by simply multiplying the PFDs of each relief valve ($0.01 \times 0.01 = 0.0001$). Even though each valve has an individual PFD = 0.01, the IPL data table for the combination of two redundant relief valves of similar design suggests a generic PFD value of 0.001 to account for the potential for common cause failure. To use a lower PFD for a dual pressure relief valve system, a thorough evaluation of the equipment, process conditions, and management systems should be performed to ensure that the potential for common cause failure is sufficiently managed.

3.3   FUNCTIONALITY

For an IPL to be effective, it needs to *function* in a way that prevents or mitigates the consequence of the scenario being studied. An IPL is required to perform its intended function under the actual process operating conditions during the event, and a number of factors are considered to ensure that the IPL will operate effectively. Some of these factors are listed below:

- The IPL design basis applies to the particular scenario for which it is credited. Consideration of standards and practices (such as those from NFPA, ASME, API, etc.) can help ensure that recommended safeguards qualify as IPLs.
- The IPL is valid for the mode of operation being analyzed (startup, shutdown, normal, batch, etc.).
- When operator response is part of the IPL, there is a well-written procedure and an effective training program to ensure that operators understand the hazard and how to respond to the initiating event, an alarm annunciation, or an emergency situation.

- The IPL is able to accomplish its function in sufficient time to prevent the consequence of concern. In the case of a human IPL, there is enough time for the operator to return the process to a safe state.

3.3.1 Time Dependency

A critical aspect of functionality is that the response of the IPL be *timely*. When an initiating event occurs, the process condition changes from normal to abnormal operation. The process deviation can ultimately lead to the consequence of concern at a rate related to the scenario conditions and the process design. An evaluation of an IPL is important to confirm that the IPL can successfully complete its action and that the process can return to a safe operating condition within the *process safety time* (PST). As defined in *Guidelines for Safe and Reliable Instrumented Protective Systems* (CCPS 2007b), the PST is

> *"The time period between a failure occurring in the process or its control system and the occurrence of the hazardous event."*

The PST represents an overall scenario timeline. IPLs step in along this timeline and take action. The time interval between IPL operations can affect the capability of an IPL to perform in the intended sequence. For example, if a SIS setpoint is too close to the PRV set pressure, the PRV might lift before the SIS completes its action. Similarly, a human acting as an IPL needs to be able to detect, diagnose, and respond in sufficient time to prevent the consequence of concern and before further troubleshooting in the field becomes unsafe.

In performing a LOPA, it is important to consider how the IPLs operate and to ensure that there are no conflicting states of operation or improper sequencing. It is possible that unsafe states may occur if some IPLs operate and others do not. Just as a failure of one IPL should not affect the operation of another, the early or delayed operation of an IPL should not cause a functional failure of another. Each IPL is expected to be fully functional and independent so that it is capable of stopping the scenario progression without human action or the operation of any other device, including other IPLs.

Each safeguard credited as an IPL in LOPA effectively executes its function faster than the process condition degrades, thus preventing the ultimate consequence of concern. To ensure that the IPL will perform its intended function, two important time-based parameters can be considered: the *IPL response time* (IRT) and the *process lag time* (PLT). The IRT is the time for a given IPL to detect an out-of-limit condition and complete the actions intended to achieve a safe state (but does not include the time of full recovery to the safe state). The IRT is dependent on the specific IPL design and may change as the design is modified. Once the IPL has successfully responded, the PLT represents how much time it will take for the process to achieve or maintain a safe state. The PLT is a function of the specific process being protected. It is critical to consider both the IRT and the PLT when designing an IPL.

Example: To protect against high pressure in a column, a proposed IPL will turn off the heat to the column's thermosiphon reboiler. It is possible to design an IPL with a fast IRT that can detect the column pressure and close a heating valve to the reboiler very quickly. However, the liquid and metal components in the reboiler remain hot and may represent a heat source to the process for a much longer period of time. Even though the IPL can respond quickly, due to the extended process lag time inherent in the system, this IPL may actually not be fast enough to prevent column overpressure.

By understanding both the IRT and the PLT of the system, it is possible to determine the *maximum setpoint* (MSP) for each IPL. The MSP represents the maximum deviation from the normal operating process conditions that would still allow sufficient time for the IPL to detect the abnormal condition, for the IPL to respond, and for the process to achieve or maintain a safe state. If the deviation is detected before the process condition deviation reaches the MSP, there is sufficient time for the IPL to take action and for the process to respond. If the deviation is detected beyond the MSP, then there will be insufficient time to successfully bring the process to a safe state. The process should not exceed its design limit because necessary action is taken at the MSP. It is recommended that the MSP (or the "never exceed, never deviate" point) include a safety factor. The amount of safety factor used would be dependent on the dynamics and variability of the process. The larger the variability, the larger the recommended safety factor should be.

It is important to recognize that the PLT is not a fixed parameter. It is a dynamic function of a process based on the changing process conditions along the timeline of the scenario. In the case of a runaway reaction, setting the IPL setpoint at a low temperature may allow for many hours of response time. If the IPL setpoint is set at a higher temperature, there may be significantly less time available for a response. Figure 3.1 provides a simple illustration of the timeline for the progression of a scenario. It should be noted that the relationship between the degree of deviation from normal operating conditions and the resulting time available to successfully respond is generally not linear for a runaway reaction scenario.

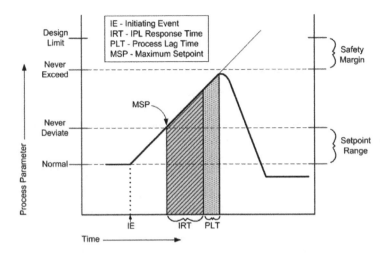

Figure 3.1. Timeline for progression of a scenario.

It is important that the LOPA team has a fundamental understanding of the process and of the operation of the IPL to be able to properly select a MSP that will allow sufficient time for detection, response, and system recovery. The team may choose to be conservative in selecting an IPL setpoint, considering such factors as potential lag in detection due to varying process conditions or component wear over time. IRT, PLT, and MSP are interrelated. If one parameter is changed, the impact on the other two should be considered. For example, if the LOPA team decides to implement a BPCS IPL but later in the project decides to change this to an alarm IPL (requiring a much longer IRT for an operator to perform the same action), the alarm setting may need to be correspondingly closer to the initial deviation. In general, it is recommended that MSP for alarm IPLs be set as close to the normal operating limit as practical.

IPLs that act to prevent the consequence of concern are covered by specifications that consider how quickly each IPL is required to detect and respond to a process deviation. The specification determines at what point the IPL initiates its action and how much time the IPL has to function. For example, a scenario may take several hours to propagate from the initiating event to the consequence of concern, but the operator may only have a short time to respond if the safety alarm is not initiated until late in the incident scenario.

IPLs that act after the loss event has occurred to reduce the harm caused by the event are also specified to detect and respond to the incident in a timely manner. With post-release IPLs, the intent is generally to initiate action as quickly as feasible and to continue providing the function until the incident has ended. For example, if an emergency ventilation system is actuated upon detection of a toxic vapor release, the ventilation system should be evaluated to determine if it can adequately dissipate or dilute the released vapors from the time of release through the time required to evacuate people from the area.

3.3.2   SCAI and Response Time

Safety controls, alarms, and interlocks (SCAI) are often relied upon to identify abnormal operation and report it to the operator or to take action to achieve or maintain a safe state for the process. The available time is limited by the method used to detect the abnormal condition, the speed at which the condition changes, and the level of process deviation reached before action is taken. Each of these limitations has uncertainty that should be assessed during the design phase.

Careful evaluation during the LOPA will ensure that the selected MSP provides sufficient time for the SCAI to complete its response and for the process to return to a safe operating condition. For many SCAI, the instrument action is much faster than the progression of the scenario. However, even a fast SIS can fail to protect if it is initiated too late in the scenario sequence. In some scenarios, more than one SCAI may be implemented to reduce the risk. These SCAI may be specified to operate in a defined sequence through MSP selection, instrumented system design, or process design.

As an example, consider a high pressure scenario that relies upon a SIS and a rupture disk. The SIS sensor(s) and the rupture disk have specified actuation points for operation. If the SIS MSP is set too close to the rupture disk opening pressure, the rupture disk could potentially open prior to the SIS taking action. It is also possible for the SIS to take action prior to the rupture disk opening, but the process response to its action is not fast enough to prevent the rupture disk from opening. While the risk from the scenario under study may be addressed by this design, the opening of the rupture disk creates another scenario. The new scenario may be undesirable and may also be preventable by the SIS if a lower MSP is selected.

The time required to execute a particular SCAI sequence is highly dependent on the application and the specific system design. For example, changes in pressure can be detected in seconds, while an analyzer may take minutes to report a process condition. Actions on the process may be taken by final control elements, such as valves, that may take seconds or hours to complete their action, depending on their design specification.

Guidelines for Safe and Reliable Instrumented Protective Systems (CCPS 2007b) suggests that SCAI be designed to respond and complete its action within 50% of the allocated process safety time (or required response time). This safety factor is intended to ensure that the automated systems can complete their defined actions given the uncertainty associated with the various lags, measurement errors, and accuracy issues that occur within these systems. ISA TR84.00.04 (2011) provides guidance on determining the SIS MSP.

3.3.3   Human-Based IPLs and Response Time

A human-based IPL may be initiated by an alarm, field reading, sample analysis, or other action demanded at a specific point along a scenario timeline. Human reliability is significantly impacted by the amount of time available to detect, diagnose, and respond to a scenario. This human response time should be less than the time it takes for the consequence of concern to occur from the point of detection. The time available for the individual to respond is limited by when in the scenario timeline the human is able to detect the problem. For example, if the IPL is dependent on an operator responding to an alarm, the human response time is limited by how long it takes for the scenario to propagate to the consequence of concern from the alarm MSP. The overall response time, from scenario detection through the process transitioning toward a safe state, should be less than the time it would take for the consequence of concern to occur if no action is taken in response to the alarm. Note that the instrumented systems involved in the detection of the scenario and in responding to the scenario are considered SCAI. Refer to Section 5.2.2.1 for SCAI requirements.

When evaluating the effectiveness of the alarm IPL, consideration is given to the sequential operation of any IPL after the alarm IPL is initiated. For an alarm response, the time before a subsequent safety shutdown or pressure relief valve (PRV) activation should be sufficient to allow the operator to complete the specified action.

In the *Handbook of Human Reliability Analysis with Emphasis on Nuclear Power Plant Applications* (Swain and Guttmann 1983), estimates were provided for the probability of human failure to respond correctly in the time following an alarm annunciation. Figure 3.2 is extracted from the *Handbook* to reflect part of the response by control room personnel to an abnormal event. This graph represents the probability that an operator will fail to properly diagnose a situation after a compelling alarm if no other alarms occur. The error probabilities given include both the probability of the operator failing to detect the alarm and the probability of the operator failing to diagnose the situation correctly.

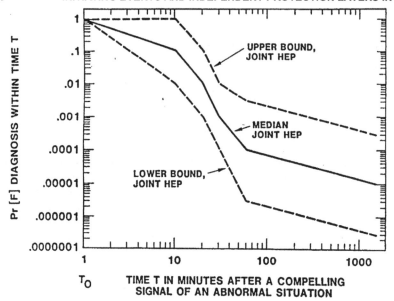

Figure 3.2. Joint probability of error of detection and diagnosis, versus the time available for detection and diagnosis (Swain and Guttmann 1983).

In this figure,

HEP = Human error probability

Pr [F] = Probability of failure (in this case, the probability of an operator failing to detect and diagnose correctly)

Note from Figure 3.2 that the more time available to detect and diagnose the problem, the lower the chance of error.

Figure 3.2 only covers part of the incident timeline. After diagnosis, the operator is required to respond appropriately to prevent the scenario from propagating to the next protection layer or to the ultimate consequence. The response time for an alarm IPL is

$$IRT_{alarm} = t_{Alarm} + t_{Operator} + t_{Process}$$

where

IRT_{alarm} is the IPL response time for the alarm.

t_{Alarm} is the time for the alarm system to annunciate the alarm condition.

$t_{Operator}$ is the time for the operator to detect, diagnose, and complete the proper response steps.

$t_{Process}$ is the time for the process to achieve a safe operating condition.

In summary, the response time required for a human IPL is influenced by the following:

1. If based on an alarm, the time for the sensor to detect the critical limit and announce this to the worker (annunciation time)
2. The time for the human to detect/perceive the alarm or other indication of an abnormal situation (detection time)
3. The time for the worker to decide on a course of action (decision time)
4. The time to diagnose the problem; steps can include determining if the alarm is false (diagnosis time)
5. The time for the worker to complete the required action, including returning the process to normal condition, placing the process in a safe state, or shutting the process down (action time)
6. The time required for the process to return to a safe state once the IPL action is completed (process lag time)

If the diagnosis or response requires that the worker enter the area(s) potentially affected by the hazard scenario, the analyst should also consider whether the process safety time is adequate for the worker to confirm the success (or failure) of his actions and safely exit the area. If not, the response may not be credited as a human IPL.

Delays and lag time can occur at any point along the scenario timeline. For example, the operator may not be present at the operator station when the alarm is first annunciated, or the operator may be required to walk several minutes to access the equipment. If the overall detection and response, including any delay in the annunciation prompting the response, can be accomplished within the time required, the human response meets the time constraint of an IPL. (The other criteria of an IPL should also be met, as described in Chapter 5.)

3.4   INTEGRITY

An understanding of the relative magnitude of risk reduction achievable by a specific IPL is essential to determine whether there is sufficient protection against the consequence of concern. *Integrity* is a property of the IPL that is a measure of its capability to satisfy its specified requirements. The integrity claim for any IPL is constrained by the level of risk reduction that the IPL can reasonably achieve as designed, installed, operated, and maintained in the operating environment. Integrity is also influenced by the procedures and practices used by an organization to minimize the likelihood of human error that could lead to IPL failure. Ultimately, the dependability of both equipment and human IPLs is limited by the strength of the management systems that support their ongoing effectiveness.

For an IPL to be dependable, it must be available when needed. The ability to detect and correct failures in a timely manner reduces the amount of time that the process operates under upset conditions, in transient operations (such as maintenance or shutdown), or without the protection of a fully functional IPL. If a failure is immediately apparent to the operator through observation, indication, or alarm, corrective action can be taken within a relatively short period of time. The degree to which the failure is revealed affects the availability of the IPL and increases the integrity of the system.

3.4.1   Integrity of Equipment

The integrity of any IPL is limited by its weakest element. Equipment integrity is affected by many factors, including the hardware and software design, the operating environment, the ITPM program, and the management system used to prevent systematic error.

Some IPLs are simple devices, such as a restrictive orifice. For simple IPLs, the IPL integrity can be estimated by considering historical performance and examining equipment records. Data tables associated with simple devices often include specific design and ITPM recommendations.

In contrast, other IPLs can be quite complex and require multiple components to operate correctly in order to achieve the required functionality and risk reduction. Estimating the performance of complex systems often requires additional quantitative analysis. For example, SISs are relatively complex systems that are specific to the process and are configured to accomplish specific functions related to the identified hazard scenarios. SISs can be sufficiently complex such that IEC 61511 (2003) Clause 11.9 requires that the risk reduction be verified using quantitative techniques. For complex IPLs, the data tables often suggest additional analysis rather than providing specific recommendations.

3.4.2   Integrity as Related to Human IPLs

The integrity of an IPL that relies on human decisions and actions is limited by the ability of the operator to be able to detect, diagnose, and act to stop the propagation of the scenario. Human performance is dependent on quality data, accurate information, and effective interfaces. Consequently, the evaluation of human IPLs generally includes consideration of the human factors associated with equipment, procedures, and the operating environment, as they influence the human's ability to effectively perform the appropriate response action in a timely manner. In Chapter 5, the human IPL PFD values shown in the data tables are generally based on the assumption that the response can be performed within the stated response time. Estimation of the response time can be qualitative or can be based on more specialized techniques. For some IPLs, simulations, drills, or trial runs are recommended to verify that the estimated response time is achievable.

3.4.3   Revealed versus Unrevealed Failure

Many equipment failures are obvious when they occur. Failures that initiate LOPA scenarios are typically considered to be *revealed* failures because they result in an identifiable process deviation. For example, a normally operating pump failing off can be revealed by a loss of flow, or a control valve unexpectedly closing during material transfer could reveal a failure of the level control loop.

Some IPL failures can be revealed during normal operation. If an IPL operates as part of normal process control, the failure of the IPL may be detectable prior to it being required to stop a scenario of concern. For example, consider a scenario in which the initiating event is the failure of a flow totalizer into a vessel and the consequence of concern is vessel overflow. The vessel is equipped with a level indicator that is independent of the flow totalizer, and this level indicator is part of an IPL to protect against high level in the vessel. If the level indicator malfunctions, failure of this vessel level IPL could be revealed during normal operation based on the operator's calculation of what is transferred in and out of the tank. Some IPLs are also equipped with diagnostics that indicate when the IPL is not functioning properly. IPL may be equipped with fault detection and alarms that alert operating personnel to a malfunction. This permits timely detection and correction of the IPL failure prior to it being challenged during an event.

The term *unrevealed* applies to IPLs that do not operate as part of normal process operation, so whether or not they are in a failed state is not generally known until there is a process deviation. An example of an unrevealed failure is the malfunction of a high-high level switch in a tank without fault diagnostics. If the switch fails, the tank will continue to operate normally until an initiating cause occurs and the failed high-high level switch does not prevent the tank from overfilling. Because IPL failure is not generally observable by the operator through process upset, failures of IPLs are prevented by an asset integrity plan that incorporates inspection, proof testing, and validation to ensure their reliability.

The use of diagnostic equipment to reveal a failure would not be considered an IPL in LOPA. However, the ability to detect equipment failure does reduce the amount of time that any system is out of service or operating in an unintended state. The use of diagnostics increases the reliability of the system and may result in a lower PFD for a particular IPL.

The data tables found in Chapters 4 and 5 contain limited design and ITPM guidance intended to support equipment integrity and reliability. IPLs are often addressed in detail in other codes, standards, and practices; this document is not intended to replace those documents. Further, detailed assessment of the IPL design and associated asset integrity program is outside the scope of LOPA. The guidance is intended to clarify when certain types of design or validation activities are assumed as part of the claimed IPL PFD value.

3.5 RELIABILITY

Guidelines for Safe and Reliable Instrumented Protective Systems (CCPS 2007b) provided the following definition of *reliability*:

> *"Reliability is an attribute of a protection layer related to its equipment operating as intended, under stated conditions, for a specified time period."*

A holistic understanding of the IPL design, the chemical process that the IPL protects, the IPL's intended function within that process, and the environment in which the IPL will operate is needed to determine the required reliability of an IPL. An IPL is considered reliable when it operates as required, when required, does not experience frequent out-of-service periods, and does not act partially or completely without cause. Reliability is therefore related to the IPL performing as intended in the operating environment for an expected time.

This concept encompasses more than the theoretical probability of individual system elements operating successfully when required. For some IPLs, specific design, operation, inspection, and maintenance practices are needed to achieve sufficient IPL reliability. For example, a detonation arrester is designed for the specific process environment and subjected to rigorous testing based on defined protocols and specifications to assure reliability.

3.5.1   Low Demand Mode

In addition to ensuring that the IPL functions as anticipated, it is also important that the IPL is called upon at the frequency anticipated. In LOPA, IPLs are expected to be challenged infrequently; this is referred to as *low demand mode*. An IPL is operating in low demand mode if the device is challenged less than once per year. The simple math for an IPL in low demand mode can be illustrated for a scenario involving an initiating event with one IPL.

$$f_1^C = IEF_1 \times PFD_1$$

where

f_1^C = Frequency of the consequence occurring for scenario 1

IEF_1 = Frequency of the IE for scenario 1

PFD_1 = Probability of failure on demand of independent protection layer 1 for scenario 1.

The LOPA equation assumes that the layers are sufficiently independent such that the potential for a common cause failure between the initiating event and IPLs is low compared to the estimated IEF.

3.5.2   High Demand Mode

When the IPL is challenged more than once a year, the IPL is considered to be operating in *high demand mode*. In this case, the method for calculating the scenario frequency is different from that used for an IPL in low demand mode.

An example is the case of a backup generator that acts as an IPL and provides power in the event of a plant power outage. If the plant loses power 10 times per year, the generator is required to function at a high demand rate. In such cases, the analyst should use the failure of the generator to start as the IE and use the corresponding IEF value for the generator failure (e.g., IEF=1/10 years) instead.

Looking at this on the basis of the f_i equation,

f_I^C = IEF_I x PFD_I = 10 demand/yr x 0.1/demand = 1 event/yr; however, this is greater than the estimated frequency of the generator failing to start

The actual IEF is

f_I^C = 1/10 years that the generator fails to start

ISA TR84.00.04 Annex I (2011) contains several examples related to SIS and mode of operation. As an example, consider a batch process produces an off-gas containing hydrocarbons (see Figure 3.3). The off-gas is sent to a catalytic incinerator to destroy the hydrocarbons prior to the off-gas being vented to the atmosphere. High levels of hydrocarbons in the off-gas can cause excessive heating in the catalytic incinerator and potential damage to the catalyst. A SIS is implemented to detect high levels of hydrocarbons, take action to isolate the steam, and flood the header with nitrogen. The SIS was assigned a SIL 2 in the process hazard analysis. The planned SIS design has a dangerous failure rate of 1/200 years (0.005/yr) and achieves SIL 2 performance at annual testing (PFD=0.0025). The batch process experiences abnormal operation and is shut down about once every 3 to 4 batches by the SIS. Assuming that 100 batches are produced annually, the SIS operates approximately 25 to 35 times per year. This is high demand mode operation, since the SIS is activated more frequently than once per year.

Looking at this on the basis of the f_i^C equation,

f_i^C = IEF_i x PFD_{il} = 35 demand/yr x 0.0025/demand = 0.085 events/yr

The calculated f_i^C (0.085/yr) is greater than the SIS dangerous failure rate (0.005/yr); this is not possible, from a mathematical perspective.

The previous example from ISA TR84.00.04 (2011) illustrates the use of actual (rather than rounded, order of magnitude) numbers in LOPA. Using order of magnitude LOPA, the results illustrate even more clearly the mathematical impossibility using the low demand calculation:

f_i^C = IEF_i x PFD_{il} = 100 demand/yr x 0.01/demand = 1 event/yr

The calculated f_i^C (1/yr) is greater than the SIS dangerous failure rate (0.005/yr).

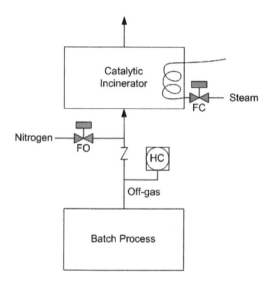

Figure 3.3. Example of a SIS in high demand mode.

In both cases, the math is predicting an event frequency that is higher than can actually be expected. The hazardous event frequency is not given by the number of demands per year; rather, it is limited by the SIS dangerous failure rate. The SIS is operating as a control function in this example, and the process hazard analysis should be changed to reflect SIS failure as the initiating cause. Alternatively, consideration should be given to modifying the process control scheme to reduce the demand rate so that the SIS operates in low demand.

3.6   AUDITABILITY

Auditability reflects the ability of an organization to inspect procedures, records, previous validation assessments, and other documented information to ensure that design, testing, maintenance, and operation continue to conform to expectations. It is necessary to have a means to audit the performance of management systems related to IEs and IPLs. When IPLs are being identified by a LOPA team, it is recommended that consideration be given to the means by which the proposed IPL could be audited to ensure ongoing effectiveness.

- Maintenance systems are periodically audited to verify that the existing administrative processes ensure that maintenance is performed as required, and that design, maintenance, and proof test documentation is maintained.

- Auditing assures that the organizational requirements for validation, procedures, and training are being followed, and that the required results are being achieved.

- The MOC process is audited to ensure that changes to materials, operating parameters, equipment, procedures, and organization are properly reviewed and documented, and that any action points generated in the review have been completed.

- Auditing cycles are set at an appropriate frequency to ensure that management systems remain robust, and the expected levels of equipment and operating performance are achieved.

The LOPA approach used in an organization may also be periodically audited to verify that the IEF and IPL PFD values being used in the risk assessment agree with historical performance. As new data become available, LOPAs can be revalidated to ensure that the claimed IEF and IPL PFD values are in alignment with achievable performance. Where the values used are not appropriate, the LOPA scenario can be updated and recommendations can be made to address any identified gaps.

3.7 ACCESS SECURITY

Access security includes the use of physical and/or administrative controls to reduce the chances of unauthorized system changes which may impair a safety device. The chemical process industry employs a number of means to reduce the chances of unauthorized system changes being made. Several examples are listed below:

- *Security over the BPCS* – Sites depend on security measures to prevent inadvertent or unapproved modifications of the BPCS. An incident scenario can result from a programmer entering an incorrect BPCS setting, so having a system for managing changes to the BPCS is important to reduce the chance of such an error. It is also important to restrict local and remote system access to prevent unauthorized access, bypasses, setpoint adjustment, operating mode, or programming changes. Security measures can also protect against viruses and cyber attacks.

- *Locks and Car Seal Procedures* – These are systems intended to ensure that valves or other safety devices are maintained in a defined state by means of a management system. A typical car seal management system includes the following:

  1. A list of car sealed valves and/or devices

  2. A documented basis for each car sealed position or status

  3. The position of car sealed valves and devices are checked on a regular basis

  4. Periodic, independent field audits to confirm that the positions of the car sealed valves are correct

5. Periodic audits of the inspection documentation to ensure that routine inspections are being performed as per organizational requirements. Incidents of car sealed valves or devices being in the wrong position are recorded, and corrective actions are taken.

Site-specific data gathered from an inspection program will allow the effectiveness of the car seal program to be evaluated.

• *Captive Key/Lock Systems* – Captive key/lock systems are intended to ensure that valves, other systems, or devices are operated in a defined sequence. A captive key is released once the valve or switch is placed in the proper position. The key can then be inserted in another device to change its position or operational state.

• *Limit Switches on Bypass Valves or Isolation Valves* – These can be used to detect if a valve or switch is placed in the wrong position; this can prevent isolation of a safety device.

These are only four examples of methods available to control and monitor access security. Assessing potential security breaches and determining the risks associated with those breaches will allow an organization to define necessary security measures.

3.8   MANAGEMENT OF CHANGE

Management of change (MOC) is a formal process used to review, approve, and document changes to procedures, materials, processes, equipment, or facilities. Modifications to the process or mode of operation, such as changes in raw materials, processing conditions, or equipment, may create new LOPA scenarios or reduce the effectiveness of existing IPLs. All changes to the process, operation, or maintenance practices should be assessed to ensure that previously identified LOPA scenarios are still valid, to identify new scenarios created by the change, and to ensure that existing protection layers are not compromised.

Process changes can be voluntary; for example, the system may be debottlenecked for increased productivity. Changes can also be involuntary; for example, obsolete equipment may fail and be replaced by a different model because the original model is no longer available. For IPLs, change can include taking equipment out of service (bypassing) for testing, or temporarily operating with an impaired IPL. Management of change reviews with appropriate approvals will ensure that compensating measures are in place that provide risk reduction equivalent to that provided by the operational IPL. Without proper precautions and compensating measures, the process would be in a state of higher risk than the LOPA had estimated.

Any change to a procedure, schedule, method, component, software, or material should be controlled. Changes to processes, people, and organizations should also be controlled. This control process includes

- Identifying the change and its technical basis
- Conducting risk reviews and alternative risk control planning
- Approving the change
- Specifying limitations on time periods for temporary changes and bypasses
- Documenting changes and risk reviews
- Validating of the quality of change implementation
- Updating of related documents such as procedures, records, schedules, and parts list
- Training and communication of changes to affected workers
- Ensuring that adequate and competent personnel are available to maintain IPLs

It is critical that changes that could impact the frequency of IEs or the PFDs of IPLs be carefully managed to ensure the continuing safety of the operation.

3.9   USE OF DATA TABLES

Inherent in the values in the data tables provided in Chapters 4 and 5 is the assumption that the management systems discussed in Chapter 2 are in place. The management systems will help ensure that the core attributes described in this chapter have been incorporated during system design and remain in place for the lifecycle of the system. In addition, all of the criteria listed for a candidate IE or IPL should be met before the suggested IEF or IPL PFD value is used for LOPA. When the word "must" is used, the *Guidelines* subcommittee feels that the IEF or IPL PFD value is only valid if those criteria are met. If the word "should" is used, then an alternative protocol can be used to achieve the same criteria/goal. The IEF and IPL PFD values shown are considered reasonable by this *Guidelines* subcommittee based on the subcommittee members' extensive experience in applying LOPA. Alternative values (including higher and lower IEFs or PFDs) may be used by an organization if the organization has determined that the revised values can be achieved and maintained. One method of achieving a better IEF or IPL PFD value is to follow the guidance provided in the site-specific validation methods discussed in Appendices B and C and in the data tables provided in Chapters 4 and 5.

REFERENCES

CCPS. 1998a. *Guidelines for Improving Plant Reliability through Data Collection and Analysis.* New York: AIChE.

CCPS. 2000. *Guidelines for Chemical Process Quantitative Risk Analysis,* 2nd Edition. New York: AIChE.

CCPS. 2007b. *Guidelines for Safe and Reliable Instrumented Protective Systems.* New York: AIChE.

IEC (International Electrotechnical Commission). 2003. *Functional Safety: Safety Instrumented Systems for the Process Industry Sector – Part 1: Framework, Definitions, System, Hardware and Software Requirements.* IEC 61511. Geneva: IEC.

ISA (International Society of Automation). 2002. *Safety Instrumented Functions (SIF) – Safety Integrity Level (SIL) Evaluation Techniques.* TR84.00.02-2002. Research Triangle Park, NC: ISA.

ISA. 2011. *Guidelines for the Implementation of ANSI/ISA 84.00.01.* TR84.00.04-2011 Part 1. Research Triangle Park, NC: ISA.

Swain, A., and H. Guttmann. 1983. *Handbook of Human Reliability Analysis with Emphasis on Nuclear Power Plant Applications.* NUREG CR-1278. Washington, DC: U.S. Nuclear Regulatory Commission.

4

EXAMPLE INITIATING EVENTS AND IE FREQUENCIES

4.1   OVERVIEW OF INITIATING EVENTS

An initiating event (IE) is a device failure, system failure, external event, or improper human action that begins a sequence of events leading to one or more undesired consequences of definable severity.

An IE includes failures such as

- Control loop malfunctions
- Operator skipping a step in a startup procedure
- Pump stopping unexpectedly

In some cases, the IE can include integrity failures, such as pipeline breaches and tank wall failures, that result in a loss-of-containment event. Organizations define the endpoint consequences of their LOPA scenarios based on their specific risk criteria; some use loss of containment as the final consequence of concern, whereas other organizations may use endpoint consequences such as personal injury, fatality, environmental impact, or economic impact. In cases of equipment integrity failure, where loss of containment of material is the consequence of the LOPA scenario (rather than a specific impact arising from that release), there may be no viable IPLs that will prevent the actual loss of containment. As a result, LOPA may not be the appropriate tool to use to evaluate equipment integrity failure scenarios where loss of containment is the defined consequence of concern. (For example, LOPA may not be the best tool to assess of the adequacy of an ITPM program to prevent loss of containment from piping as a result of corrosion under insulation.) Loss of containment IEs are further discussed in Section 4.3.4.

The unreliability of the device, system, human action, or the frequency (likelihood) of an external event affects the risk of a hazard scenario. The lower the frequency (i.e., less likely to occur), the lower the risk is to the organization from that scenario. Initiating events are always expressed as frequencies (number of occurrences over a specified time frame); for LOPA, the frequencies are usually expressed as events per year, often on an order of magnitude basis.

The reliability of equipment that could initiate a LOPA scenario is dependent upon a number of factors that are influenced by the management systems discussed in Chapter 2. These include proper equipment specification and design, the processing environment, the operating conditions, and the effectiveness of the ITPM program. Periodic management system audits are important to confirm that effective management systems are in place to ensure that equipment and human performance are maintained at sufficiently high levels.

The IEF for an IE is best determined using actual, site-specific failure rate or error rate data. However, such data may not be available or the data may not be extensive enough. In such cases, the LOPA analyst may instead use other sources of data, such as the company's LOPA protocol, expert judgment, generic industry data, or predicted IEFs. The IEFs presented later in this chapter represent consensus values from industry representatives. Some of the IEFs given in the data tables have their original basis in site-specific data, some are based on generic industry data, and some are based on expert opinion.

4.2   INHERENTLY SAFER DESIGN AND INITIATING EVENT FREQUENCY

Inherently safer design is a fundamentally different approach to reducing the risk associated with chemical processes. *Inherently Safer Chemical Processes: A Life Cycle Approach,* 2nd Edition (CCPS 2009b) has defined inherently safer design as

> *"A way of thinking about the design of chemical processes and plants that focuses on the elimination or reduction of hazards, rather than on their management or control."*

If an inherently safer design approach is used, it is possible that an initiating event is no longer plausible, and the scenario can be eliminated from the LOPA. The following are some specific examples of inherently safer designs and how to consider them during LOPA:

- If a toxic or flammable material can be removed from the process, LOPA scenarios related to release of those materials can be eliminated.
- If hazardous intermediates are consumed as they are generated rather than stored, scenarios associated with the storage of the intermediates are eliminated.

- If a pump is designed to have a maximum discharge head that is lower than the MAWP of a connected vessel or piping, the likelihood of piping or vessel rupture due to deadheading of the pump would generally be considered to have a negligible contribution to the risk; therefore, this scenario could be eliminated.

Application of inherently safer design techniques to reduce or eliminate hazards can be done during any stage of the process lifecycle. Refer to *Inherently Safer Chemical Processes: A Life Cycle Approach,* 2nd Edition (CCPS 2009b) for more guidance on the use of inherently safer design principles.

4.3   SPECIFIC INITIATING EVENTS FOR USE IN LOPA

Provided in the rest of this chapter are brief descriptions of candidate IEs with general descriptions of each IE category and illustrative examples. In some cases, the generic IEFs are based on research from within a company represented on this *Guidelines* subcommittee. In all cases, the *Guidelines* subcommittee reached consensus on the values provided in each data table. Also mentioned are IEs for which the *Guidelines* subcommittee could not offer a generic value or that may be best analyzed using a method other than LOPA. Chapter 6 discusses some of these in brief; however, the LOPA analyst may need to use other quantitative risk methods to evaluate scenarios involving IEs not listed here and for which generic data are not available.

The IE data tables in this chapter provide a brief description of each candidate IE, a typical generic IEF for the IE, and criteria associated with each value. These IEF values are generally considered to be conservative values under the conditions specified. If, however, a site has data indicating that the actual IEF is higher than that shown in the data table, the actual site value should be used. It also is possible that site-specific data collection or the use of a more quantitative risk assessment may indicate that a lower IEF may be used. Appendices B and C provide additional guidance on the development of site-specific equipment and human performance data. Each site should ensure that the IEF values selected are applicable to their facility.

Many companies, including some that participated in the preparation of this document, use IEFs that are not listed in Chapter 4. See Section 4.5 for more discussion on this issue and for the general situation of, "What If Your Candidate IE Is Not Shown in a Data Table?"

CHAPTER 4 DATA TABLES

4.3.1   Instrumented System Initiating Events

The process industry relies on instrumented systems to control the process within the normal operating limits and maintain a safe state. The failure of these control systems is one of the most cited IEs in LOPA. Since control functions are typically implemented in standard industrial equipment, these systems are referred to as basic process control systems (BPCS). Historically, the control functions have been referred to as control loops, so this text uses BPCS control loop to describe the equipment used to accomplish a specific control function. The functions acting in response to the IE are instrumented safeguards and are defined as safety controls, alarms, and interlocks (SCAI) by ANSI/ISA 84.91.01 (2012). Safety instrumented systems (SIS) are a type of SCAI that are designed and managed according to IEC 61511 (2003). The control and safeguarding functions may be performed automatically or by an operator who supervises the process and, when necessary, takes action on the process using its instrumented systems.

Malfunction in a SCAI can initiate a scenario of concern in two ways:

1.   A BPCS control loop failure may result in a process deviation that leads to a scenario of concern.
2.   Spurious operation of SCAI may result in abnormal operation, especially when partial activation occurs.

These two types of initiating events are described in Data Tables 4.1 and 4.2. For additional information about SCAI, refer to Section 5.2.2.1.

BPCS Control Loop Failure

Description: A BPCS control loop is executed by a pneumatic, electrical, electronic, or programmable electronic system (PES). It is used to regulate or guide the operation of a machine, apparatus, process, or system. A BPCS control loop is commonly implemented using pneumatic controllers, programmable logic controllers (PLC), distributed control systems (DCS), discrete controllers (nonprogrammable electronic), and single loop controllers (SLC). These controls may be safety instrumented functions if they are designed and managed in accordance with IEC 61511 (2003).

Consequences Caused: The process parameter controlled by the BPCS control loop deviates without the ability to recover on its own, resulting in a consequence of concern.

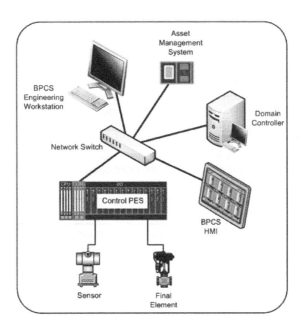

Figure 4.1. Example of BPCS control loop using a PES (Summers 2013).

Data Table 4.1. BPCS control loop failure

Initiating event description
BPCS control loop failure
Generic IEF suggested for use in LOPA
0.1/yr
Special considerations for use of generic IEF for this IE
• Instrumentation and controls that normally operate to support process (or regulatory) control fail, initiating the scenario progression. These controls may be safety instrumented systems (SIS), if they are designed and managed in accordance with IEC 61511 (2003). • The dangerous failure rate of a BPCS (which does not conform to IEC 61511 [2003]) that places a demand on a protection layer shall not be assumed to be < 10^{-5}/hr (Clause 8.2.2), which is approximately 0.1/yr.
Initial quality assurance
Initial validation of performance during commissioning.
Generic validation method
• The schedule for the ITPM task depends on the reliability desired (such as 0.1/yr failure rate) and the site experience of what scheduled or condition-based tasks are necessary to achieve 0.1/yr or better. • The failure of a BPCS control loop is generally revealed through process operation, usually when compared with local process indicators (pressure gauges, sight glasses, or other process variable measurements) or with trends of upstream or downstream indicators. • Repair of the system is initiated when failure occurs, is detected, or when calibration checks/diagnostics indicate incipient conditions. • The IE frequency can be verified by tracking historical performance.
Starting source of guidance
Consensus of the *Guidelines* subcommittee and ANSI/ISA 84.00.01-2004 Part 1 (IEC 61511-1 Mod) (ANSI/ISA 2004), Clause 8.2.2, specifies a failure rate no lower than 10^{-5}/hr.

Spurious Operation of SCAI

Description: SCAI are instrumented safeguards that take action to achieve or maintain a safe state of the process in response to a specified process condition. Refer to Section 5.2.2.1 for more detailed information on SCAI. When judging the risk reduction capability of SCAI, the system is assessed based on the potential that it experiences a dangerous failure that results in the loop failing to operate when the demand occurs. SCAI can also exhibit safe failures, where the SCAI may partially or completely activate even though the process conditions do not warrant its activation. These safe failures are classified in industry database taxonomies as spurious operation or nuisance trip (ISA TR84.00.02 2002).

Complete activation results in SCAI taking its specified action. For example, the spurious operation of a sensor initiates the safety alarm without cause. The sensor failure may later be classified as safe, but the operator must decide whether to take action on the process or to simply report the sensor failure. Complete activation normally does not result in the initiation of a chain of events leading to the specific scenario the SCAI was designed to stop. However, it is important that the LOPA team understands how the action of the SCAI affects the process operation. For example, in the event of high level, a SIS may initiate closure of an inlet valve to prevent overfilling of the tank. This action may prevent tank overflow; however, the valve closure may also block the pipeline, resulting in high pressure upstream. Whenever SCAI are added as safeguards, they reduce the risk of the hazard scenarios that they are designed to address, but they can contribute to process upsets and initiate other scenarios of concern.

Partial activation can cause process upsets by triggering a state of operation that is neither normal nor completely shut down. For example, a SIS that isolates multiple pipelines can experience an output failure on a controller that results in a single spurious valve closure and the loss of supply from a single process pipeline. The partial activation may contribute to scenarios already considered in LOPA or may result in new scenarios not previously considered. It is important to assess the impact of spurious action of each final element to ensure that any hazard scenarios created are considered in the LOPA.

Advanced quantitative techniques can be used to estimate the spurious failure rate of SCAI. When a spurious trip rate is predicted by this method, the value can be used in place of the generic value given in Data Table 4.2. It is recommended that the potential for systematic and human errors also be considered when selecting a final value. Site-specific data can be used for this IEF when sufficient process operating history exists.

Consequences Caused: The spurious operation of SCAI may lead to an upset or other consequences of concern.

Data Table 4.2. Spurious operation of SCAI

Initiating event description
Spurious operation of SCAI
Generic IEF suggested for use in LOPA
The value may be derived from calculation based on the design (typically ranges from 1/yr to 0.1/yr).
Special considerations for use of generic IEF for this IE
This IEF covers the complete system, including the sensor(s), logic solver(s), and final element(s).
Initial quality assurance
The initial validation of the system performance is confirmed during normal commissioning.
Generic validation method
• Preventive maintenance is conducted according to the manufacturer's recommendations. • Incipient conditions that could result in spurious or premature operation of the system may be detected during preventive maintenance activities. • The premature operation of the system is revealed through process operation. • Repair of the system is initiated when failure occurs. • The IE frequency used in LOPA can be verified by tracking historical performance.
Starting source of guidance
Consensus of the *Guidelines* subcommittee and calculations of spurious trip rates for SCAI.

4.3.2   Human Error Initiating Events

Human error is associated with a significant percentage of incidents in the chemical industry. As such, it is important to consider scenarios initiated by human error in LOPA. To use the generic human error IEFs presented in the subsequent data tables, the following are expected to be in place:

- A written procedure is available describing the required actions.
- The operator is trained on the procedure.
- The failure rate of the equipment components is low enough such that the combined equipment and human error rates meet the stated IEF.
- Human factors are reasonably controlled. Refer to Appendix A for more information on factors that influence human error rate.

The frequency at which an error can be expected is a function of both the predicted error rate per task and how many times the task is performed within a given time period. Useful references for calculating human error rates include *Handbook of Human Reliability Analysis with Emphasis on Nuclear Power Plant Applications,* NUREG CR-1278 (Swain and Guttmann 1983) and *Human Reliability and Safety Analysis Data Handbook* (Gertman and Blackman 1994).

For ease of use in order of magnitude LOPA, the three data tables presented in this section provide IEFs on an annualized basis associated with the following task frequencies:

1. The task is performed at least weekly.
2. The task is performed between once a week and once a month.
3. The task is performed less often than once a month.

The less often a task is performed, the fewer opportunities an individual has to make a mistake. Thus, the annual error rate would tend to decrease as the task frequency decreases. However, when a task is performed less often, the individual is typically less skilled in the task, tending to increase human error rate. When performing nonroutine work (where a given individual performs a task less than once a month), it is a good idea for the individual to review the procedure prior to doing the task or make use of a checklist or other job aid to minimize the risk of human error. The *Guidelines* subcommittee was not able to generate a suggested generic IEF value for complex tasks. It is suggested that companies select failure rate data based upon site-specific data as described in Appendix B.

Human Error during a Routine Task That Is Performed ≥ Once per Week

Description: A human error occurs on a task that is performed at a frequency of once per week or more often.

Consequences Caused: The consequences are dependent on the task being performed by the person.

Data Table 4.3. Human error during a routine task that is performed ≥ once per week

Initiating event description
Human error during a routine task that is performed once per week or more often
Generic IEF suggested for use in LOPA
1/yr
Special considerations for use of generic IEF for this IE
A procedure that is completed frequently gives the person performing the task the opportunity to develop a higher level of skill; this would tend to lower the error rate. However, the impact of this enhanced skill level is more than offset by the greater number of opportunities that the person has to make an error when the task is performed frequently. Also, people can become complacent when a task becomes routine; this can lead to higher error rates as well. The error rate value provided in this data table can be used as the IE frequency for an error in completing a task more than once a week that results in a consequence of concern. For the generic value to be applicable, • The task is low in complexity, with step-by-step instructions. • The task is documented in a procedure. • The person performing the task is trained on the procedure. • Human factors have been reasonably controlled. (See Appendix A for a discussion of human factors that need to be managed.)
Initial quality assurance
An initial training plan with specific requirements for detailed review of the procedure and demonstrated skill by the worker in performing the required task has been completed.
Generic validation method
• Procedure update and training systems are in place and periodic reviews indicate that procedures and training are up to date. • Procedures are up to date and those performing the task have received training. • Available human error data on performance of the task are reviewed.
Starting source of guidance
Consensus of the *Guidelines* subcommittee. For Generic IEF and validation, NUREG CR-1278 *Handbook of Human Reliability Analysis with Emphasis on Nuclear Power Plant Applications, Final Report* (Swain and Guttmann 1983) provides a variety of human error data. See Appendix A for a list of performance shaping factors and their relative influences on human error rates.

Human Error during a Task That Is Performed between Once per Month and Once per Week

Description: This initiating event is a human error in overlooking a step or performing a step incorrectly on a task that is performed less than once per week but at least once per month.

Consequences Caused: The consequences are dependent upon the task being performed by the person.

Data Table 4.4. Human error during a task that is performed between once per month and once per week

Initiating event description
Human error during a task that is performed between once per month and once per week
Generic IEF suggested for use in LOPA
0.1/yr
Special considerations for use of generic IEF for this IE
A procedure that is done at a moderate frequency helps to maintain skill level after training and has a lower risk of the complacency that can affect tasks done very frequently. There are fewer opportunities for a human error to occur during in this frequency range as compared to a procedure that is completed more frequently than once a week. For the generic value to apply, • The task is low in complexity, with step-by-step instructions. • The task is documented in a procedure. • The person performing the task is trained on the procedure. Note, if a specific individual does not perform the given task at least once per month, the task is considered to be non-routine for that individual. In this case, it is recommended that the individual review the procedure before the task is performed or use another performance aid, such as a checklist. • Human factors have been reasonably controlled; refer to Appendix A.
Initial quality assurance
An initial training plan, with specific requirements for detailed review of the procedure and demonstrated skill by the worker in performing the required task, has been completed.
Generic validation method
• Procedure update and training systems are in place, and periodic reviews indicate that procedures and training are up to date. • If the task is non-routine for certain individuals, it is recommended that procedures be reviewed prior to completing the task or that performance aids, such as checklists, be used. • Available human error data on performance of the task are reviewed.
Starting source of guidance
Consensus of the *Guidelines* subcommittee. For Generic IEF and validation, NUREG CR-1278 *Handbook of Human Reliability Analysis with Emphasis on Nuclear Power Plant Applications, Final Report* (Swain and Guttmann 1983) provides human error data. Refer to Appendix A for more information.

Human Error during a Non-Routine Task That Is Performed < Once per Month

Description: This initiating event is a human error in overlooking a step or performing a step incorrectly on a task that is performed infrequently, less than once per month.

Consequences Caused: The consequences are dependent upon the task being performed by the person.

Data Table 4.5. Human error during a non-routine task that is performed < once per month

Initiating event description
Human error during a non-routine task that is performed less than once per month
Generic IEF suggested for use in LOPA
0.01/yr
Special considerations for use of generic IEF for this IE
When a task is performed less frequently there are fewer opportunities for an error. However, operators can be less skilled in the execution of infrequent, non-routine procedures. To maintain an acceptable skill level such that the generic value is applicable, • The task is low in complexity, with step-by-step instructions. • The task is documented in a procedure. • The person performing the task is trained on the procedure with refresher training every 3 years. • The individual reviews the procedure before the task is performed or uses another performance aid, such as a checklist. • Human factors have been reasonably controlled. (See Appendix A for a discussion of human factors that should be managed.)
Initial quality assurance
An initial training plan, with specific requirements for training and demonstration of these required actions, has been completed.
Generic validation method
• Procedure update and training systems are in place and periodic reviews indicate that procedures and training are up to date. • There is verification that, if the task is non-routine for certain individuals, procedures are being reviewed prior to completing the task or that performance aids, such as checklists, are being used. • Available human error data on performance of the task are reviewed.
Starting source of guidance
Consensus of the *Guidelines* subcommittee. For Generic IEF and validation, NUREG CR-1278 *Handbook of Human Reliability Analysis with Emphasis on Nuclear Power Plant Applications, Final Report* (Swain and Guttmann 1983) provides a variety of human error data. See Appendix A for a list of performance shaping factors and their relative influences on human error rates.

4.3.3   Active Mechanical Component Initiating Events

Many LOPA scenarios begin with a trigger event resulting from failure of an active mechanical component that creates a process control deviation, resulting in a consequence of concern. The following data tables describe common active mechanical component IEs used in LOPA.

Pressure Regulator Failure

Description: This data table covers the failure of the pressure regulating device to control pressure as designed. This may occur due to issues such as fouling or blockage, spring failure, diaphragm rupture, or damage to the seat. Pressure regulators may fail to open, fail to close, or fail to control at the proper pressure, resulting in a consequence of concern.

Maintenance practices vary widely, and can significantly affect the reliability of pressure regulators. With good ITPM practices, an improved failure rate may be achieved.

Consequences Caused: Loss of pressure control could cause process upsets due to high or low pressures and/or flows, depending on the process failure mechanism.

Data Table 4.6. Pressure regulator failure

Initiating event description
Pressure regulator failure

Generic IEF suggested for use in LOPA
0.1/yr

Special considerations for use of generic IEF for this IE
This scenario covers a self-contained pressure regulator in pressure reducing or backpressure service, operating in continuous control mode, which fails to operate as designed (open or closed).

Initial quality assurance
The proper operation of the pressure regulator is confirmed during normal commissioning.

Generic validation method
• The failure of a pressure regulator is revealed through process operation, usually by comparison to local process indicators (pressure gauges or sight glasses) or trends of upstream or downstream indicators. • Repair of the system is initiated when failure occurs and is detected, or when calibration checks/diagnostics indicate incipient conditions. Repair history is recorded.

Starting source of guidance
Consensus of the *Guidelines* subcommittee and *Failure Rates – Reliability Physics* (Earles and Eddins 1962), as referenced in NPRD-95 (RIAC1995), lists generic failure rate of 0.1/yr.

Screw Conveyor Failure

Description: A screw conveyor is a mechanical device that can move solid material at varying speeds over straight distances. The drive shaft is supported by bearings at the trough ends, and a motor is mounted at the discharge end. Failure of the bearings, motor, or other components can occur, resulting in the screw conveyor stopping prematurely/spuriously. Note, if the conveyor stops frequently, the first IPL to respond to the process deviation may be in high demand mode. In this case, a different IEF may be appropriate for the scenario. Refer to Chapter 3 for information regarding high demand mode.

Consequences Caused: The failure of the screw conveyor stops the process flow, resulting in an upstream and/or downstream upset or other consequence of concern.

Data Table 4.7. Screw conveyor failure

Initiating event description
Screw conveyor failure
Generic IEF suggested for use in LOPA
1/yr to 10/yr
Special considerations for use of generic IEF for this IE
• The failure rate of a conveyor system is dependent on the initial design, the corrosive or erosive nature of the material that is being conveyed, and the failure rate of the individual components of the drive train (such as the electrical motor, including controls, bearings, coupling, screw shaft, belts, etc.). The IE frequency range is typical; however, the failure rate is dependent on the physical characteristics of the material being conveyed and the processing conditions. Therefore, the IE can be higher. • The equipment is designed to withstand complete blockage of the screw conveyor and the maximum torque that the motor can apply to the conveyor.
Initial quality assurance
The conveyor moving parts are installed such that the alignment is in accordance with the manufacturer's recommendation. Initial proof testing of the performance capability is performed during commissioning. The bearings are installed correctly, lubricated, and sealed for the site environment. B20.1-2012 *Safety Standard for Conveyors and Related Equipment* (ANSI/ASME 2012) is a reference for conveyors and related equipment. ANSI/CEMA 350-2009, *Screw Conveyors for Bulk Materials* (ANSI/CEMA 2008) is a source of good engineering practices and dimensional standards.
Generic validation method
• A preventive maintenance program is implemented for the electrical motor and controls based on the plant environment and conditions. • Maintenance intervals are set as recommended by the manufacturer and based on past operating experience.
Starting source of guidance
Consensus of the *Guidelines* subcommittee. Also refer to B20.1-2012 *Safety Standard Conveyors and Related Equipment* (ANSI/ASME 2012) and ANSI/CEMA 350-2009, *Screw Conveyors for Bulk Materials* (ANSI/CEMA 2008).

Screw Conveyor Overheating of Materials

Description: There are a number of failure modes by which overheating of materials in a screw conveyor can occur. The motor can overheat due to excessive torque being applied. A screw conveyor bearing can overheat due to lack of lubrication or material entering the bearings. Frictional heating of the conveyed material can also occur as a result of foreign material (tramp metal) entering the conveyor or misalignment of the screw, causing the screw to rub against the housing/ barrel. These failures can result in thermal instability of the material being conveyed.

The frequency of material overheating is influenced by the material being conveyed and the processing conditions; therefore, it may be necessary to increase the IEF if overheating is experienced more frequently than the generic value provided in the data table would indicate.

Consequences Caused: Overheating of the conveyed material, potentially resulting in ignition or decomposition of material within the conveyor.

Data Table 4.8. Screw conveyor overheating of materials

Initiating event description
Screw conveyor overheating of materials
Generic IEF suggested for use in LOPA
0.1/yr
Special considerations for use of generic IEF for this IE
• It is recommended that the maximum torque of the conveyor drive motor be limited to prevent overheating of the conveyed material. • Safeguards typically include an overload trip on the motor driving the screw, a temperature sensor in the conveyor trough/barrel that automatically trips the motor and/or activates a water deluge system or snuffing steam, and the use of magnets to remove tramp ferrous metals.
Initial quality assurance
• The conveyor's moving parts are installed such that the alignment is in accordance with the manufacturer's recommendation. • Initial proof testing of the performance capability is performed during commissioning. • The bearings are installed correctly, lubricated, and sealed for the site environment. • B20.1-2012 *Safety Standard for Conveyors and Related Equipment* (ANSI/ASME 2012) is a reference for conveyors and related equipment. ANSI/CEMA 350-2009, *Screw Conveyors for Bulk Materials* (ANSI/CEMA 2008) is a source of good engineering practices and dimensional standards.
Generic validation method
• Maintenance is performed as recommended by the manufacturer and based on previous inspection history. • Inspections can identify incipient failures of conveyor wear surfaces, seals, and bearings. Alignment of the drive screw and housing/barrel is checked to prevent overheating. • Inspections of the conveyor are conducted to detect product buildup and signs of decomposition products.
Starting source of guidance
Consensus of the *Guidelines* subcommittee. Also refer to B20.1-2012 *Safety Standard for Conveyors and Related Equipment* (ASME 2012) and ANSI/CEMA 350-2009, *Screw Conveyors for Bulk Materials* (ANSI/CEMA 2008).

Pump, Compressor, Fan, or Blower Failure

Description: An electric- or steam-powered pump, compressor, fan, or blower fails off due to causes such as local start-station failure, shaft/coupling failure, motor failure, or impeller failure. Fans and blowers are generally used to convey air or other process gases. Failure modes include vibration, insufficient bearing lubrication or overheating, and excessive stress leading to fatigue cracking. Fans in erosive, corrosive, or high temperature service are particularly susceptible to failure. For good reliability, it is important to ensure that fans are properly mounted and aligned, bearings are periodically lubricated, and the fan is operated in the process environment for which it was designed.

Consequences Caused: This loss of operation could result in process upset, with a number of possible consequences as a result of process deviation.

Data Table 4.9. Pump, compressor, fan, or blower failure

Initiating event description
Pump, compressor, fan, or blower failure
Generic IEF suggested for use in LOPA
0.1/yr
Special considerations for use of generic IEF for this IE
The IEF applies to the complete or partial loss of function of this process component. This IEF does not include failures due to loss of power. (Refer to Data Table 4.10 for the IEF for localized power interruption.)
Initial quality assurance
Initial validation of performance is completed during commissioning, including validation of materials of construction and monitoring of vibration, bearing temperature, output pressure, and flow.
Generic validation method
The failure of a pump, compressor, fan, or blower is revealed through process operation, usually with comparison to local process indicators (pressure gauges, sight glasses, or flow indicators), or by comparison to trends of upstream or downstream indicators. • Repair of the system is initiated when failure occurs and is detected, or when walk-through checks and/or diagnostics indicate incipient conditions. • Repairs and maintenance are documented. • Maintenance intervals are set as recommended by the manufacturer and based on past operating experience.
Starting source of guidance
Consensus of the *Guidelines* subcommittee.

Site-Wide Loss of Power

Description: Loss of power occurs in a site/process due to any cause.

Loss of power is a common initiating event for LOPA scenarios. Power loss can result in the failure of electrically energized devices, including process equipment and process control instrumentation. However, site experience with power failures can vary widely. Some sites rarely lose power, some sites lose power several times each year due to weather events such as hurricanes and snow storms, and some sites can experience rolling blackouts as often as daily during high power usage periods. Therefore, a generic value for the IEF of loss of power cannot be determined by the *Guidelines* subcommittee. If loss of power occurs frequently, there might be high demand upon the first IPL. In this case, a different IEF may be appropriate for the scenario. Refer to Chapter 3 for information regarding high demand mode.

Localized Loss of Power

Description: Loss of power occurs in a single circuit. This can result from a number of possible component failures. Examples of such components include circuit breakers, switches, relays, semiconductors, wire, cables, and wiring connections. These failures can result in a localized loss of power to a specific piece of equipment or multiple pieces of equipment connected to a common power supply circuit.

Consequences Caused: Deactivation of electrically energized equipment, including controls not connected to an emergency power supply, results in a number of potential consequences.

Data Table 4.10. Localized loss of power

Initiating event description
Single circuit loss of power
Generic IEF suggested for use in LOPA
0.1/yr
Special considerations for use of generic IEF for this IE
This IEF applies to the complete or partial loss of local power due to a component failure in single circuit. This value does not include the frequency of site-wide power loss, which is site-specific.
Initial quality assurance
Initial validation of performance is conducted during commissioning of all electrical supply components.
Generic validation method
The loss of power is revealed through disruption in process operations. • Repair of the system is initiated when failure occurs and is detected, or when walk-through checks and/or diagnostics indicate incipient conditions. • Maintenance and repairs are documented.
Starting source of guidance
Consensus of the *Guidelines* subcommittee. Table 3-3 in IEEE Standard 493 (2007) provides failure rates for typical electrical components in a simple radial system at 480V point of use.

Check Valves

Check valves are mechanical devices installed in piping that permit flow in one direction but impede flow in the reverse direction. There are a variety of internal mechanisms to prevent backflow, such as balls, flaps, springs, and disks. For additional information on types of check valves, refer to Section 5.2.2.3.

Generally, check valves are considered in the classification of IPLs. There are certain applications, however, where check valves actually act as control devices (such as to prevent backflow during compressor recycling). In these situations, the check valve could be acting in high demand, and failure of the check valve could be an initiating event for a scenario. (Refer to Section 3.5.2 for more discussion on high demand mode.)

Following are IEFs for a single check valve and two check valves operating in series as control devices.

Single Check Valve Failure – *For a scenario where the check valve is in high demand mode.*

Description: The failure of a check valve to close on demand results in sufficient reverse flow to cause the consequence of concern. This entry represents a single check valve that is challenged more frequently than once a year. If the check valve is in batch service and is tested before each use, then its failure is not an IE; instead, the check valve is a candidate IPL.

Consequences Caused: The consequence that could occur as a result of reverse flow is dependent on the specific process and is generally identified in the process hazard analysis.

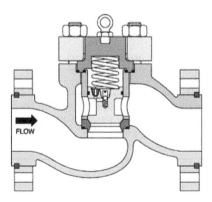

Figure 4.2. Piston check valve, courtesy of Flowserve Corporation.

Data Table 4.11. Single check valve failure

Initiating event description
Single check valve failure
Generic IEF suggested for use in LOPA
0.1/yr
Special considerations for use of generic IEF for this IE
The process service is assumed to be clean vapor, steam, water, or another non-plugging service. This is an uncommon situation where the check valve is used as a control device, and the IEF is being provided is for the anticipated failure rate as a control device. ITPM programs for check valves have not traditionally been as robust as for other devices; therefore, the analyst is cautioned to ensure appropriate ITPM practices are in place prior to using the IEF shown in this data table.
Initial quality assurance
• Verification of the material of construction for the check valve is done before installation. • Verification of the direction of the check valve is done after installation. • An operational readiness review before startup verifies the proper installation.
Generic validation method
• Routine preventive maintenance is performed at an interval based upon the severity of service and on past inspection history. • Testing can be performed in situ or off-line by back-pressuring. Internal inspection can indicate the onset of failure from mechanisms such as fouling, plugging, sticking, or corrosion. • Maintenance and repair records are maintained.
Starting source of guidance
Consensus of the *Guidelines* subcommittee and *Reliability Data Book for Components in Swedish Nuclear Power Plants*, RKS/SKI 85-25, p 79 (Bento et al. 1987). The data were then converted to IEF and PFD values using FTA. See Appendix D for additional details.

Failure of Double Check Valves in Series – *This data table pertains to a scenario where, if one out of two (1oo2) check valves works, the consequence is prevented. This IE applies when the check valves are in high demand mode.*

Description: The failure of a double check valve system to close on demand results in sufficient reverse flow to cause the consequence of concern. This data table pertains to a double check valve system that is challenged more frequently than once per year. As such, the system is operating in high demand mode. Instead of acting as an IPL, the double check valve system should be treated as an initiating event, and the failure rate of the double check valve system should be used. Refer to Section 3.5.2 for additional information on high demand mode.

Check valves are designed to operate based on differential pressure, and there needs to be sufficient differential pressure in the system to properly operate both check valves. It is also important to functionally test each check valve individually to be able to take credit for a double check valve system. One of the check valves could be in a failed state for an extended period of time (i.e., an unrevealed failure), and the dual check valve system would still pass a pressure test if the system is tested as a single unit. Testing of each check valve individually may not always be practical; however, if both check valves are not individually tested, the IEF for a single check valve would be applicable.

When using redundant systems, consideration should be given to the potential for common cause failure. The valves may be the same model or similar design, and therefore subject to similar risks associated with improper design, manufacturing defects, or incorrect material of construction. Both valves are also exposed to the same process conditions; this could create plugging or fouling that can affect both check valves. The likelihood of common cause failure can be reduced by using check valves of different types, performing maintenance on the two valves at different intervals, and inspecting at a frequency sufficient to ensure that the valves are not operating in a plugging or fouling service.

Consequences Caused: The consequence of reverse flow is dependent on the process and is generally identified as a consequence of concern in the process hazard analysis.

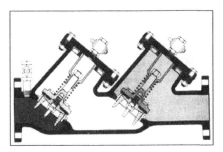

Figure 4.3. Double check valve backflow prevention device (U.S. EPA 2003).

Data Table 4.12. Failure of double check valves in series

Initiating event description
Failure of double check valves in series
Generic IEF suggested for use in LOPA
0.01/yr
Special considerations for use of generic IEF for this IE
• The process service is assumed to be clean vapor, steam, water, or another non-plugging service. This is an unusual situation where the check valve is used as a control device, and the IE value being provided is for the anticipated failure rate of the check valve when it is being used as a control device. • ITPM programs for check valves have historically not been as robust as for other devices; therefore, the analyst is cautioned to ensure appropriate ITPM practices are in place prior to using the IEF shown in this data table.
Initial quality assurance
• Verification of the material of construction of the check valve is done before installation. • Verification of direction of the check valve is done after installation. • An operational readiness review before startup verifies the proper installation.
Generic validation method
• Routine preventive maintenance is performed at an interval based upon the severity of service and past inspection history. • Testing can be performed in situ or off-line by back-pressuring. Internal inspection can indicate the onset of failure from mechanisms such as fouling, plugging, sticking, or corrosion. • Maintenance and repair records are maintained.
Starting source of guidance
Consensus of the *Guidelines* subcommittee and *Reliability Data Book for Components in Swedish Nuclear Power Plants*, RKS/SKI 85-25, p 79 (Bento et al.1987). The data were then converted to IEF and IPL PFD values using FTA. See Appendix D for additional details.

Cooling Water Failure

CCPS LOPA (2001) provided an example IEF range to represent cooling water failure. However, cooling water failure is actually a consequence resulting from one or more initiating events. Each of these discrete IEs could have a different initiating event frequency and could potentially require different IPLs for risk management. For example, loss of cooling could result from a human error, such as leaving a valve closed. It could result from a failure of a cooling pump or associated piping. It could also result from a control loop failure. These individual causes of loss of cooling, with their associated generic IEF values, are discussed in the data tables shown in this chapter for the specific IEs of interest. It is recommended that cooling water failure scenarios be analyzed by examining each of the specific initiating events that could cause loss of cooling.

4.3.4   Loss of Containment Initiating Events

Loss of containment events in the process industries can have significant consequences as a result of hazardous material and energy releases. Loss of containment events are some of the most frequently evaluated hazard scenarios in LOPA. Some loss of containment IEs lead directly to a release, and there are no identified IPLs able to prevent the release from occurring. In these cases, the only potential IPLs for such events may be those that limit the impact of the event. Some common mitigative layers of protection (such as containment and firefighting systems) may be effective in some scenarios. In other scenarios, the required mitigative systems can be complex, and the assessment of these systems requires the use of advanced consequence modeling and quantitative risk assessment techniques. Complex mitigative layers are often specific to a particular site, so many of these are not included in the scope of this document.

Some common initiating events that can lead to a release of material are discussed in Section 4.3.4.1. Other, less frequent loss of containment initiating events, such as piping leaks and vessel ruptures, are discussed in Section 4.3.4.2.

4.3.4.1 Common Loss of Containment Initiating Events

Primary Pump Seal Leak

Description: Some pumps use external drives (electric motors, steam turbines, expanders, etc.) that are coupled to the pump housing to provide power. The pump shaft passes through a sealing mechanism to allow the pump to operate without loss of the fluid to the environment. This sealing mechanism may be a simple packed gland, a single mechanical seal, or a complex multi-seal system using two or more seal systems (with the space between the seals pressurized to prevent leakage). Seal selection is based on a number of factors including temperature, pressure, rotational speed, and the properties of the material contacting the seal.

All simple single seal systems normally leak at some rate. The normal leak rates are very low and are sometimes measured as part of fugitive emission monitoring programs. The degree of leak tolerance from a particular seal is a key factor in pump seal selection.

Pump seal leak rates of process safety concern are those that would pose a hazard to workers or create conditions where a fire/explosion could occur. Such releases occur when a pump seal component fails or is worn to the point where it no longer functions. Common leak points in a pump seal are the insert mounting, gland gasket, and shaft packing.

Causes of pump leaks include both design and installation factors and environmental factors. Design and installation issues include poor selection of material of construction, misalignment, and bearing failure.

Environmental factors that can adversely impact pump seal life include hot service, erosive service such as slurries, corrosive service, high pressure service, and flashing service. It is important to select materials of construction that are resistant to all fluids that may be pumped as well as any cleaners or solvents that may be used to flush through the lines. Another factor that has a significant impact on seal life is the proximity of the pump's normal operating point to the best efficiency point (BEP) on the pump operation curve. A pump operating close to BEP (75–115%) will have significantly longer seal life than a pump operating far above or far below this range.

Some applications use a mechanical seal system that includes more than one seal. This design includes a primary and a secondary (or outboard) seal with the inter-seal space under pressure or vented to a safe place. For either arrangement, a leak by the primary seal will either allow the seal fluid to leak into the process stream or allow the leaking process fluid to be vented to a safe location, with no loss of containment to the surroundings. Many design variations have been developed; refer to API 682 (2002) for more information. Designs of this type allow for a monitoring system of the inter-seal area so that a primary seal leak can be detected.

The IEF suggested by the *Guidelines* committee applies to the PRIMARY seal, which is the same whether the configuration is a single or dual mechanical seal. (As with any IEF, a site may substantiate a better value with sufficient data; refer to Appendix C for additional guidance on site-specific equipment failure rate data collection.) The outboard, secondary seal may not have the same integrity as the inner, primary seal and failure may be accelerated if the barrier fluid becomes contaminated or runs dry following the inner seal failure. Therefore, a dual mechanical seal system without leak detection *does not* necessarily reduce the risk of a loss of containment by an order of magnitude. The use of a secondary pump seal with effective leak detection system and response is considered to be an IPL protecting against failure of the primary seal and is covered in Chapter 5.

When considering a LOPA scenario involving a pump seal leak, it is important to understand the magnitude of the consequence of concern. Seal leaks generally begin by showing seepage. Where there is the potential for more serious consequences, such as a major fire with potential for fatality, a robust inspection program that can detect the onset of leakage significantly reduces the probability of a large release scenario. Understanding the type of failure is necessary to create the consequence of concern is key to selecting the appropriate IEF.

Consequences Caused: Typically, the consequence of a pump seal leak is a minor, continuous release of the pump's material through the primary seal to the environment, which can sometimes increase over time if not corrected.

Data Table 4.13. Pump seal leak

Initiating event description
Pump seal leak
Generic IEF suggested for use in LOPA
1/yr
Special considerations for use of generic IEF for this IE
• Properly sizing the pump for the application will increase the life of the seal. • Pump foundation design, piping design and installation, and pump alignment can all significantly impact the seal life on a pump.
Initial quality assurance
Cartridge seals are leak tested prior to installation.
Generic validation method
• Visual inspections can be documented during routine operator rounds, included in inspection checklists, or recorded in databases. • Fugitive emissions monitoring is typically conducted for environmental regulatory compliance where required. • Maintenance and operating experience support the failure rate used.
Starting source of guidance
Consensus of the *Guidelines* subcommittee.

Sealless Pump Failure (Loss of Containment)

Description: Pumps that do not have seals as pressure boundaries fall into this category. Canned pumps and magnetic drive couplings are examples of these pumps, which could represent inherently more reliable solutions for preventing loss of containment. IE frequencies with respect to loss of containment of the process fluid from sealless pumps are likely to be one to three orders of magnitude better than pumps with seals. However, the actual failure rate would be highly dependent upon the process fluid and operating conditions, so a generic failure rate cannot be recommended by the *Guidelines* subcommittee. It should be recognized that sealless pumps are often selected when the process fluid is very aggressive. The combination of process fluid, process conditions, pump design, and materials of construction will strongly influence the failure rate, and actual operating experience is a better source of data. Although sealless pumps may have a lower overall failure rate with regard to loss of containment, due to the aggressive nature of the process, the use of sealless pumps may also create other design and operability issues that also need to be assessed and managed.

Complete Primary Pump Seal Failure

Description: A release occurs due to total mechanical failure of the primary pump seal (and secondary pump seal, if present). Catastrophic failure of one or more O-rings in a mechanical seal can result in a significant release of the fluid being pumped through the resulting annular orifice into the environment. An identification of the mechanism of the pump seal failure will aid in defining effective IPLs that can halt the progression of the event.

Consequences Caused: Typically, a large, continuous release of process material occurs through the pump seal(s) to the environment.

Alternative Design Option: Disaster bushings are long, narrow openings between the pump casing and shaft that limit the rate of release in the event of a catastrophic seal failure. Typically, the annular opening of a disaster bushing is a few thousandths of an inch wide. The consequence of a seal failure is minimized significantly due to the small area available for leakage.

Data Table 4.14. Complete primary pump seal failure

Initiating event description
Complete primary pump seal failure
Generic IEF suggested for use in LOPA
0.1/yr
Special considerations for use of generic IEF for this IE
This scenario reflects the frequency of complete failure of a primary pump seal. The actual failure frequency may differ from the generic value in this data table, depending on service and the robustness of the bearing/bearing lube system.
Initial quality assurance
Cartridge seals are leak tested prior to installation.
Generic validation method
• Visual inspections can be documented during routine operator rounds, on inspection checklists, or recorded in databases. • Fugitive emissions monitoring is typically conducted for environmental regulatory compliance, and documentation of testing is maintained for the life of the application or as specified by the applicable regulation. • Maintenance and operating experience support the failure rate used.
Starting source of guidance
Consensus of the *Guidelines* subcommittee.

Hose Failure, Leak and Rupture

Description: A release can occur from a hose and connected equipment as a result of hose failure. The hose is assumed to be an appropriate type for the specific service (typically braided steel or other material with sufficient integrity for the application). Exposing the hose to high levels of vibration can increase the failure rate. In some services, particularly higher hazard services, some companies replace hoses on a higher, more conservative frequency to reduce the likelihood of a hose failure.

Consequences Caused: There is a release of the contents of the hose as well as the contents of nonisolated upstream and downstream process equipment.

Data Table 4.15. Hose failure, leak and rupture

Initiating event description
Hose failure, leak and rupture
Generic IEF suggested for use in LOPA
0.1/yr for leak and 0.01/yr for rupture
Special considerations for use of generic IEF for this IE
This scenario applies to leaks or complete hose failure due to age, external damage, wear, etc. (See the human error rates in Data Tables 4.3 – 4.5 for hose releases due to poor hose connections made by individuals.) • The best hose performance is achieved by minimizing stress on the hose. • In heavily stressed services, where hoses are exposed to significant vibration, pressure cycling, bending, dragging, or wear, the generic value may not apply.
Initial quality assurance
Materials of construction are verified and the hose is inspected prior to use.
Generic validation method
• The hose is visually inspected before each use for cuts, cracks, abrasion, exposed reinforcement, stiffness or hardness of the hose, changes in color, blistered cover, kinks or flattened hose, leakage, or damaged reinforcement. • Pressure testing before each use may be necessary for some applications.
Starting source of guidance
Consensus of the *Guidelines* subcommittee and *Guidelines for Process Equipment Reliability Data with Data Tables* (CCPS 1989), p 187. The mean catastrophic rupture frequency is $0.57/10^6$ hr (or 0.005/yr).

Premature Opening of Spring-Loaded Relief Valve

Description: A spurious operation or the premature opening of a relief valve occurs, resulting in a release. (This failure mode does not include release due to leakage.)

Consequences Caused: There is a pressurized release of the process material to the environment or to a controlled discharge system.

NOTE: Spurious opening of a relief valve is much more likely to occur if the operating pressure is close to the relief valve set pressure. Setting a sufficient margin between the relief valve set pressure and the normal operating pressure of the process will reduce the frequency of spurious trips.

Data Table 4.16. Premature opening of spring-loaded relief valve

Initiating event description
Premature opening of spring-loaded relief valve
Generic IEF suggested for use in LOPA
0.01/yr
Special considerations for use of generic IEF for this IE
• Premature opening of the relief valve may lead to an upset or other consequences of concern. • The event may have safety, environmental, and economic impact. • The relief device is properly installed, and the set point is higher than the normal process operating pressure.
Initial quality assurance
The device is certified by the manufacturer or is tested by a facility certified by a responsible authority prior to installation.
Generic validation method
• The relief valve is tested by a certified shop on an appropriate frequency, based on service and past reliability. • Visual inspection can be done during plant walk-throughs; more frequent ITPM is recommended based upon process severity. • As-found and as-left conditions are documented. The relief valve is returned to the like-new condition prior to return to service.
Starting source of guidance
Consensus of the *Guidelines* subcommittee and *Guidelines for Process Equipment Reliability Data with Data Tables* (CCPS 1989), Taxonomy No. 4.3.2. Actual frequencies are reported as a range, from 0.00024/yr to 0.042/yr.

4.3.4.2 Piping and Vessel Loss-of-Containment Initiating Events

CCPS LOPA (2001) listed suggested starting frequency values for a number of loss of containment events. Since the publication of *CCPS LOPA*, some practitioners have found that LOPA may not be most appropriate method to analyze certain loss of containment events. When the consequence of concern for a particular LOPA scenario is the loss of containment of material due to an integrity failure such as a pipe or vessel leak (regardless of the actual impact of the release), it is possible that no IPLs can be identified that can prevent the release from occurring. In this case, a hazard scenario may best be addressed through proper design and material of construction selection, quality assurance, and an effective ITPM program.

When the consequence of concern relates to the impact of a loss of containment event (such as personal injury, fatality, property loss, or environmental impact), mitigative safeguards may be effective IPLs. These may include leak detection, containment, and firefighting systems. In addition, the use of conditional modifiers may be part of a company's LOPA protocol for analyzing these types of events. Refer to *Guidelines for Enabling Conditions and Conditional Modifiers in Layer of Protection Analysis* (CCPS 2013) for more guidance on the use of conditional modifiers.

Loss of containment initiating events as described in this section do not require other circumstances (such as overpressure) to result in a release. Leaks and ruptures occur randomly under normal process conditions and are generally the result of causes such as material or construction defects, fatigue, and corrosion/erosion.

The values provided in the subsequent data tables are based on data from *Guidelines for Quantitative Risk Assessment,* 2nd Edition (CPR 2005), also known as the "Purple Book." This failure rate information is gathered on an ongoing basis, and recommended failure rates may periodically change based on the availability of new data. When applying these values, it is assumed that sufficient protective measures have been taken to prevent foreseen failure mechanisms. These values are not applicable to equipment that is not covered under an effective asset management program.

The values presented in the subsequent data tables represent typical failure rates that may be expected in the chemical process industry, and they have been used as starting values for quantitative risk assessment methodologies. The choice of materials of construction can significantly impact equipment failure rates in many types of processes. Processes in which equipment is susceptible to vibration, corrosion, erosion, embrittlement, or extreme temperature ranges may require very specific materials of construction. The values presented in these data tables may not be appropriate for equipment in extreme service. Also, the values in the data tables are not recommended for assessing the failure rates of FRP vessels, tanks, and piping; the *Guidelines* subcommittee was not able to identify sufficient data and maintenance criteria to provide guidance for FRP equipment.

The failure rates suggested in the data tables reflect only the frequency of random integrity failures of the piping, tank, or vessel. These data do not include the failure rates of associated equipment, such as agitator seals, sight glasses, or instrumentation. More importantly, these values do not reflect failure rates due to equipment misoperation, such as:

- Overfilling
- Overpressuring
- Vacuum creation
- Runaway reaction
- Internal deflagration
- External damage

The frequencies of these types of failures are often higher than the frequencies of integrity failures of piping, tanks, and vessels. These types of failure events need to be evaluated in separate LOPA scenarios, in addition to considering the random containment failures discussed in this section and the subsequent data tables.

The failure rates for piping are presented on a per meter basis. To assign an appropriate IEF for piping, the system is first defined so that the length of piping associated with a particular event can be determined. The length of the piping can then be multiplied by the appropriate failure rate per meter to obtain the overall failure rate of a given system. The failure rate of a piping system can also be influenced by the number of piping connections; a leak at a gasket is more likely to occur than a leak at a welded connection.

The following data tables are included:

Data Table 4.17.   Atmospheric tank: catastrophic failure
Data Table 4.18.   Atmospheric tank: continuous 10 mm diameter leak
Data Table 4.19.   Pressure vessel: catastrophic failure
Data Table 4.20.   Aboveground piping: full breach failure (pipe size ≤ 150 mm, 6 in)
Data Table 4.21.   Aboveground piping: full breach failure (pipe size > 150 mm, 6 in)
Data Table 4.22.   Aboveground piping: leak (pipe size ≤ 150 mm, 6 in)
Data Table 4.23.   Aboveground piping: leak (pipe size > 150 mm, 6 in)

Atmospheric Tank: Catastrophic Failure

Description: The material and/or manufacturing defects inherent in equipment are the cause of the failure. The failure rate assumes that the appropriate technical specifications were used and that there is an ITPM program in place.

CCPS LOPA (2001) provided a failure frequency range of 10^{-3}/yr to 10^{-5}/yr. More recent data (CPR 2005) indicate that the random catastrophic failure rate of an atmospheric tank is 10^{-5}/yr, at the lower end of the previously suggested range. The failure rate due to equipment misoperation, such as overpressuring or overfilling, is not reflected in the suggested IEF value in Data Table 4.17. Events resulting in these consequences should be evaluated using additional LOPA scenarios.

Consequences Caused: The consequence is the immediate release of the tank contents.

Data Table 4.17. Atmospheric tank: catastrophic failure

IE description
Atmospheric tank: catastrophic failure
Generic IEF suggested for use in LOPA
0.00001/yr (or 10^{-5}/yr)
Special considerations for use of generic IEF for this IE
The IE frequency is based on a single-walled atmospheric tank in nonvibrating service that is considered fit for service based on the material of construction and the pressure rating.
Initial quality assurance
Follow the recommendations of the appropriate codes for the jurisdiction. Standards such as API 650 (2013) and API 620 (2008) (or EN 14015:2004 [BS 2005] in Europe) are useful references for design and construction.
Generic validation method
• Periodic internal/external inspection frequencies are based on specific industry standards, local regulations, and previous inspection history.
• Other techniques, such as x-ray, ultrasonic thickness measurement, magnetic particle testing, acoustic measurement, and dye penetration testing can also be used to detect incipient conditions.
• The site reviews inspection reports, modifies inspection frequencies based on inspection results, and makes repairs if there are signs of corrosion, erosion, or stress cracking.
• The site plans include remaining-life calculations based on thickness measurements from the baseline data.
Starting source of guidance
Consensus of the *Guidelines* subcommittee. Refer to *Guidelines for Quantitative Risk Assessment,* 2nd *Edition* "Purple Book" 18E (CPR 2005), p 3.6, Table 3.5, Frequencies of Loss of Containments (LOCs) for atmospheric tanks (5×10^{-6}/yr).

Atmospheric Tank: Continuous 10 mm Diameter Leak

Description: The material and/or manufacturing defects inherent in equipment are the cause of the failure. The failure rate assumes that the appropriate technical specifications were used and that there is an ITPM program in place.

Consequences Caused: The consequence is the continuous release of the tank contents from a single point source failure, assuming a 10 mm diameter hole.

Data Table 4.18. Atmospheric tank: continuous 10 mm diameter leak

IE description
Atmospheric tank: continuous 10 mm diameter leak
Generic IEF suggested for use in LOPA
0.0001/yr (or 10^{-4}/yr)
Special considerations for use of generic IEF for this IE
The IE frequency is based on a single-wall atmospheric tank in a nonvibrating service that is considered fit for service based on the material of construction and the pressure rating.
Initial quality assurance
Follow the recommendations of the appropriate codes for the jurisdiction. Standards such as API 650 (2013) and API 620 (2008) (or EIN 14015:2004 [BS 2005] in Europe) are useful references for design and construction.
Generic validation method
• Periodic internal/external inspection frequencies are based on specific industry standards, local regulations, and previous inspection history.   • Visual external and internal inspections are performed.   • Other techniques, such as x-ray, ultrasonic thickness measurement, magnetic particle testing, acoustic measurement, and dye penetration testing can also be used to detect incipient conditions.   • The site reviews inspection reports, modifies inspection frequencies based on inspection results, and makes repairs if there are signs of corrosion, erosion, or stress cracking.   • The site plans include remaining-life calculations based on thickness measurements from the baseline data.
Starting source of guidance
Consensus of the *Guidelines* subcommittee. Refer to *Guidelines for Quantitative Risk Assessment,* 2nd Edition "Purple Book" 18E (CPR 2005), p 3.6, Table 3.5, Frequencies of Loss of Containments (LOCs) for a continuous release from a 10 mm hole in an atmospheric tank directly to atmosphere (1×10^{-4}/yr).

Pressure Vessel: Catastrophic Failure

Description: The material and/or manufacturing defects inherent in equipment are the cause of the failure. The failure rate assumes that the appropriate technical specifications were used and that there is an ITPM program in place.

CCPS LOPA (2001) provided a failure frequency range for pressure vessels of 10^{-5}/yr to 10^{-7}/yr. More recent data (CPR 2005) indicate that the random catastrophic failure rate of an atmospheric tank is 10^{-5}/yr, at the higher end of the previously suggested range. Pressure vessels can be subject to a number of failure modes associated with the cycling of operating pressures and temperatures, corrosion, erosion, and vibration. These factors can influence the overall failure rate of a pressure vessel.

Consequences Caused: The consequence is the immediate release of the vessel contents during a 10 minute time period.

Data Table 4.19. Pressure vessel: catastrophic failure

IE description
Pressure vessel: catastrophic failure
Generic IEF suggested for use in LOPA
0.00001/yr (or 10^{-5}/yr)
Special considerations for use of generic IEF for this IE
The IE frequency is based on a single-walled atmospheric tank in nonvibrating service that is considered fit for duty based on the material of construction and the pressure rating.
Initial quality assurance
Follow the recommendations of the appropriate codes for the jurisdiction. Codes such as ASME Section VIII (2013) (or EN13445 *Unfired Pressure Vessels* [UNM 2002], harmonized to comply with the European *Pressure Equipment Directive 97/23/EC* [EC 1997]) are useful references for design and construction.
Generic validation method
• Periodic internal/external inspection frequencies are based on specific industry standards, local regulations, and previous inspection history. • Visual external and internal inspections are performed. • Other techniques, such as x-ray, ultrasonic thickness measurement, magnetic particle testing, acoustic measurement, and dye penetration testing can also be used to detect incipient conditions. • The site reviews inspection reports, modifies inspection frequencies based on inspection results, and makes repairs if there are signs of corrosion, erosion, or stress cracking. • The site plans include remaining-life calculations based on thickness measurements from the baseline data.
Starting source of guidance
Consensus of the *Guidelines* subcommittee. Refer to *Guidelines for Quantitative Risk Assessment,* 2nd Edition "Purple Book" 18E (CPR 2005), p 3.3, Table 3.3, Frequencies of Loss of Containments (LOCs) for stationary vessels. The value listed is 5×10^{-6}/yr for process vessels and reactors.

Underground Piping

The failure rate of underground piping can vary, depending on a number of factors. These include the pipe diameter, wall thickness, material of construction, degree of ground movement, and impact of corrosion/erosion. For instance, underground water piping of ductile iron has exhibited a failure frequency of 10^{-4} per meter per year (NRCC 1995). In comparison, the European Gas Pipeline Incident Data Group (EGIG 2011) has collected data on gas transmission pipeline systems since 1970. The average pipeline failure rate over 40 years of data collection was 3.5×10^{-7} per meter per year. This value is comparable to the failure rate of 10^{-7} per meter per year suggested in Data Table 4.21 for aboveground piping greater than 150 mm (6 inches) in diameter. The *Guidelines* subcommittee was not able to generate a suggested generic value for the failure rate of underground piping. It is suggested that companies select failure rate data based upon site-specific data or other sources of data that are representative of the piping system being evaluated.

Aboveground Piping: Full Breach Failure (Pipe Size ≤ 150 mm, 6 in)

Description: The material and/or manufacturing defects inherent in equipment are the cause of the failure. The failure rate assumes that the appropriate technical specifications were used and that there is an ITPM program in place.

Consequences Caused: The consequence is the immediate, pressurized release of the pipe contents.

Data Table 4.20. Aboveground piping: full breach failure (pipe size ≤ 150 mm, 6 in)

IE description
Aboveground piping: full breach failure (pipe size ≤ 150 mm, 6 in)
Generic IEF suggested for use in LOPA
0.000001/m of piping per year (or 10^{-6}/m of piping per year)
Special considerations for use of generic IEF for this IE
The scenario is a complete rupture of a well-supported pipe that is not in high-vibration service. The rupture is due to random failure as a result of mechanical or metallurgical issues.
Initial quality assurance
• PMI, x-ray inspection, and pressure testing are various means to ensure the initial quality of the material. • Piping specification, design, installation, and quality assurance practices follow relevant industry standards.
Generic validation method
Recommended inspection frequencies can vary based on specific industry standards, local regulations, and previous inspection history. • Periodic external inspections are conducted to look for signs of cracking or corrosion, especially under insulation. • Thickness measurements can also be used to monitor pipe integrity and detect loss of wall thickness due to corrosion or erosion. • A number of methods can be employed to monitor piping condition, such as external thickness measurements, ultrasonic testing, corrosion coupon testing, inspection under insulation for corrosion, and ultrasonic shear wave testing for stress cracking.
Starting source of guidance
Consensus of the *Guidelines* subcommittee. Refer to *Guidelines for Quantitative Risk Assessment*, 2nd Edition "Purple Book" 18E (CPR 2005), p 3.7, Table 3.7, Frequencies of Loss of Containments (LOCs) for pipes. Actual frequencies reported range from $3x10^{-7}$ to $1x10^{-6}$/m of piping per year.

Aboveground Piping: Full Breach Failure (Pipe Size > 150 mm, 6 in)

Description: The material and/or manufacturing defects inherent in equipment are the cause of the failure. The failure modes include piping failure as well as gasket failure. The failure rate assumes that the appropriate technical specifications were used and that there is an ITPM program in place.

Failure rate data indicate that the frequency of leakage from a larger pipe is at least an order of magnitude lower than the frequency of leakage from a smaller diameter pipe.

Consequences Caused: The consequence is the immediate, pressurized release of the pipe contents.

Data Table 4.21. Aboveground piping: full breach failure (pipe size > 150 mm, 6 in)

IE description
Aboveground piping in typical service: full breach failure (pipe size > 150 mm, 6 in)
Generic IEF suggested for use in LOPA
0.0000001/m of piping per year (or 10^{-7}/m of piping per year)
Special considerations for use of generic IEF for this IE
The scenario is a complete rupture of a well-supported pipe that is not in high-vibration service. The rupture is due to random failure as a result of mechanical or metallurgical issues.
Initial quality assurance
• PMI, x-ray inspection, and pressure testing levels are some methods used to ensure the initial quality of the material. • Piping specification, design, installation, and quality assurance practices follow relevant industry standards.
Generic validation method
Recommended inspection frequencies can vary based on specific industry standards, local regulations, and previous inspection history. • Periodic external inspections are conducted to look for signs of cracking or corrosion, especially under insulation. • Thickness measurements can also be used to monitor pipe integrity and detect loss of wall thickness due to corrosion or erosion. • A number of methods can be employed to monitor piping condition, such as external thickness measurements, ultrasonic testing, corrosion coupon testing, inspection under insulation for corrosion, and ultrasonic shear wave testing for stress cracking.
Starting source of guidance
Consensus of the *Guidelines* subcommittee. Refer to *Guidelines for Quantitative Risk Assessment,* 2nd Edition "Purple Book" 18E (CPR 2005), p 3.7, Table 3.7, Frequencies of LOCs for pipes. The value reported is 1×10^{-7}/m of piping per year.

Aboveground Piping: Leak (Pipe Size ≤ 150 mm, 6 in)

Description: The material and/or manufacturing defects inherent in equipment are the cause of the failure, resulting in a leak with an effective diameter of 10% of the nominal diameter (up to a maximum of 50 mm). This IE includes flange and associated gasket leaks due to mechanical or metallurgical failures. The failure modes include piping failure as well as gasket failure. The failure rate assumes that the appropriate technical specifications were used and that there is an ITPM program in place. The failure rate of a piping system is also influenced by the number of piping connections; a leak at a gasket is more likely to occur than a leak at a welded connection.

Consequences Caused: The consequence is the release of process fluid to the surroundings from the piping system leak location.

Data Table 4.22. Aboveground piping: leak (pipe size ≤ 150 mm, 6 in)

IE description
Above ground piping: leak (pipe size ≤ 150 mm, 6 in)
Generic IEF suggested for use in LOPA
0.00001/m of piping per year (or 10^{-5} /m of piping per year)
Special considerations for use of generic IEF for this IE
The scenario is a single leak of a well-supported pipe with no excessive vibration, corrosion/erosion, or thermal cyclic stresses. • The leak has an effective diameter of 10% of the nominal diameter (up to a maximum of 50 mm). • The leak is due to random failure as a result of mechanical, metallurgical issues, or gasket leaks.
Initial quality assurance
• PMI, x-ray inspection, and pressure testing levels are some methods used to ensure the initial quality of the material. • Piping specification, design, installation, and quality assurance practices follow relevant industry standards.
Generic validation method
Recommended inspection frequencies can vary based on specific industry standards, local regulations, and previous inspection history. • Periodic external inspections are conducted to look for signs of cracking or corrosion, especially under insulation. • Thickness measurements can also be used to monitor pipe integrity and detect loss of wall thickness due to corrosion or erosion. • A number of methods can be employed to monitor piping condition, such as external thickness measurements, ultrasonic testing, corrosion coupon testing, inspection under insulation for corrosion, and ultrasonic shear wave testing for stress cracking.
Starting source of guidance
Consensus of the *Guidelines* subcommittee. Refer to *Guidelines for Quantitative Risk Assessment,* 2nd Edition "Purple Book" 18E (CPR 2005), p 3.7, Table 3.7, Frequencies of LOCs for pipes. Frequencies listed range from 2×10^{-6} to 5×10^{-6}/m per year.

Aboveground Piping: Leak (Pipe Size > 150 mm, 6 in)

Description: The material and/or manufacturing defects inherent in equipment are the cause of the failure, resulting in a leak with an effective diameter of 10% of the nominal diameter (up to a maximum of 50 mm). The failure modes include piping failure as well as gasket failure. The failure rate assumes that the appropriate technical specifications were used and that there is an ITPM program in place.

Failure rate data indicate that the frequency of leakage from a larger pipe is at least an order of magnitude lower than the frequency of leakage from a smaller diameter pipe. The failure rate of a piping system is also influenced by the number of piping connections; a leak at a gasket is more likely to occur than a leak at a welded connection.

Consequences Caused: The consequence is the release of process fluid to the surroundings from the piping system leak location.

Data Table 4.23. Aboveground piping: leak (pipe size >150 mm, 6 in)

IE description
Aboveground piping: leak (pipe size >150 mm, 6 in)
Generic IEF suggested for use in LOPA
0.000001/m of piping per year (or 10^{-6}/m of piping per year)
Special considerations for use of generic IEF for this IE
The scenario is a single leak of a well-supported pipe with no excessive vibration, corrosion/erosion, or thermal cyclic stresses. • The leak has an effective diameter of 10% of the nominal diameter (up to a maximum of 50 mm). • The leak is due to random failure as a result of mechanical, metallurgical issues, or gasket leaks.
Initial quality assurance
• PMI, x-ray inspection, and pressure testing levels are some method used to ensure the initial quality of the material. • Piping specification, design, installation, and quality assurance practices follow relevant industry standards.
Generic validation method
Recommended inspection frequencies can vary based on specific industry standards, local regulations, and previous inspection history. • Periodic external inspections are conducted to look for signs of cracking or corrosion, especially under insulation. • Thickness measurements can also be used to monitor pipe integrity and detect loss of wall thickness due to corrosion or erosion. • A number of methods can be employed to monitor piping condition, such as external thickness measurements, ultrasonic testing, corrosion coupon testing, inspection under insulation for corrosion, and ultrasonic shear wave testing for stress cracking.
Starting source of guidance
Consensus of the *Guidelines* subcommittee. Refer to *Guidelines for Quantitative Risk Assessment,* 2nd Edition "Purple Book" 18E (CPR 2005), p 3.7, Table 3.7, Frequencies of LOCs for pipes. The frequency is reported as 5×10^{-7}/m of piping per year.

Boxed or Clamped Flanges

Boxed or clamped flanges are used in some industries for temporary repair of a piping leak. The probability of these temporary measures failing in a leak or rupture mode is not known, so a generic value for the IEF of a boxed or clamped flange failure could not be developed. If boxed or clamped flanges are in use, an organization can develop its own IEF based on site-specific or other appropriate data.

4.4 EXTERNAL EVENTS

External events can cause incidents and can be considered in qualitative and quantitative risk assessments. External events include

- Lightning
- Flooding
- Hurricane
- Tornado
- Earthquakes
- Land subsidence (sinkholes)
- Mudslides
- Impacts from nearby facility (knock-on effects)
- Small and large external fires that are not process related
- Airplane crash
- Other external impacts

Example values for the frequencies of small and large external fires were offered in *CCPS LOPA* (2001). However, the frequencies of fires and other external events are very variable and are based on factors such as weather, location, and other parameters highly specific to the individual facility. Therefore, it was not possible for the *Guidelines* committee to develop generic IEFs for external events.

4.5 WHAT IF YOUR CANDIDATE INITIATING EVENT IS NOT SHOWN IN A DATA TABLE?

The *Guidelines* subcommittee reviewed many candidate IEs and decided which met the criteria for IEs for LOPA. The *Guidelines* subcommittee also decided if there were sufficient data available to support listing each IE, including its generic criteria and validation approach. Many companies, including some that have participated in the writing of this document, use IEs that are not listed in Chapter 4. If a company wishes to use IEs not included in this document, the following is recommended:

- Data used to support the IEF are from a documented, reliable source or from site-specific data. Refer to Appendices B and C for more guidance on data collection.

- The general requirements for implementing and maintaining the systems that support the IEF for a given IE are met. As a best practice, companies are encouraged to publish their candidate IEs and supporting data so that others can peer review the data/criteria and potentially benefit from use of the new IE data.

REFERENCES

ANSI/ASME (American National Standards Institute/American Society of Mechanical Engineers). 2012. *Safety Standards for Conveyors and Related Equipment.* B20.1-2012. New York: ASME.

ANSI/CEMA (American National Standards Institute/Conveyor Equipment Manufacturers Association). 2008. *Screw Conveyors for Bulk Materials* 350-2009. Naples, FL: Conveyor Equipment Manufacturers Association.

ANSI/ISA (American National Standards Institute/International Society of Automation). 2004. *Functional Safety: Safety Instrumented Systems for the Process Industry Sector – Part 1: Framework, Definitions, System, Hardware and Software Requirements.* 84.00.01-2004 (IEC 61511-1 Mod). Research Triangle Park, NC: ISA.

ANSI/ISA. 2012. *Identification and Mechanical Integrity of Safety Controls, Alarms and Interlocks in the Process Industry.* ANSI/ISA-84.91.01-2012. Research Triangle Park, NC: ISA.

API (American Petroleum Institute). 2002. *Pumps - Shaft Sealing Systems for Centrifugal and Rotary Pumps*, 2nd Edition. Standard 682. Washington, DC: API.

API. 2008. *Design and Construction of Large, Welded, Low-Pressure Storage Tanks,* 11th Edition. Standard 620. Washington, DC: API.

API. 2013. *Welded Storage Tanks for Oil Storage,* 12th Edition. Standard 650. Washington, DC: API.

ASME (American Society of Mechanical Engineers). 2013. ASME Boiler and Pressure Vessel Code Section VIII – *Rules for Construction of Pressure Vessels,* Division 1. New York: ASME.

Bento J.-P., S. Björe, G. Ericsson, A. Hasler, C.-D. Lydén, L. Wallin, K. Pörn, O. Akerlund. 1987. *Reliability Data Book for Components in Swedish Nuclear Power Plants.* RKS/SKI 85-25. Stockholm: Swedish Nuclear Power Inspectorate and Nuclear Training & Safety Center of the Swedish Utilities.

BS (British Standards Institution). 2005. *Specification for the design and manufacture of site built, vertical, cylindrical, flat-bottomed, above ground, welded, steel tanks for the storage of liquids at ambient temperature and above.* EN 14015:2004 British-Adopted European Standard. London: IBN.

CCPS. 1989. *Guidelines for Process Equipment Reliability Data, with Data Tables*. New York: AIChE.

CCPS. 2001. *Layer of Protection Analysis: Simplified Process Risk Assessment*. New York: AIChE.

CCPS. 2009b. *Inherently Safer Chemical Processes: A Life Cycle Approach,* 2nd Edition. New York: AIChE.

CCPS. 2013. *Guidelines for Enabling Conditions and Conditional Modifiers in Layer of Protection Analysis*. New York: AIChE.

CPR (Committee for the Prevention of Disasters). 2005. *Guidelines for Quantitative Risk Assessment,* 2nd Edition "Purple Book." 18E. The Hague: Sdu Uitgevers.

Earles, D., and M. Eddins. 1962. *Failure Rates – Reliability Physics*. Proceedings of First Annual Symposium on the Physics of Failure in Electronics. Baltimore: Spartan Books.

EC (European Commission). 1997. *Pressure Equipment Directive 97/23/EC* (PED). Brussels: EC.

EGIG. 2011. *Gas Pipeline Incidents, 8th Report of the European Gas Pipeline Incident Data Group*. Groningen: EGIG.

Gertman, D., and H. Blackman. 1994. *Human Reliability and Safety Analysis Data Handbook*. New York: John Wiley & Sons.

IEC (International Electrotechnical Commission). 2003. *Functional Safety: Safety Instrumented Systems for the Process Industry Sector – Part 1: Framework, Definitions, System, Hardware and Software Requirements*. IEC 61511. Geneva: IEC.

IEEE. 2007. *Recommended Practice for the Design of Reliable Industrial and Commercial Power Systems*. Standard 493-2007. New York: IEEE.

ISA (International Society of Automation). 2002. *Safety Instrumented Functions (SIF) – Safety Integrity Level (SIL) Evaluation Techniques*. TR84.00.02-2002. Research Triangle Park, NC: ISA.

NRCC (National Research Council Canada). 1995. *Water Mains Breaks Data on Different Pipe Materials for 1992 and 1993*. A-7019.1. Ontario: National Research Council.

RIAC (Reliability Information Analysis Center). 1995. *Nonelectronic Parts Reliability Data*. NPRD-95. Rome, NY: RAC.

Swain, A., and H. Guttmann. 1983. *Handbook of Human Reliability Analysis with Emphasis on Nuclear Power Plant Applications*. NUREG CR-1278. Washington, DC: U.S. Nuclear Regulatory Commission.

U.S. EPA (Environmental Protection Agency). 2003. *Cross Connection Control Manual*. EPA 816-R-03-002. Washington, DC: EPA.

UNM (Union de Normalisation de la Mécanique). 2002. *Unfired Pressure Vessels*. EN 13445. Courbevoie, France: UNM.

5

EXAMPLE IPLS AND PFD VALUES

5.1 OVERVIEW OF INDEPENDENT PROTECTION LAYERS (IPLs)

An IPL is a device, system, or action that is capable of preventing a scenario from proceeding to an undesired consequence regardless of the initiating event or the action of any other protection layer associated with the scenario (*CCPS LOPA* 2001). The effectiveness of an IPL is quantified in terms of its probability of failure on demand (PFD). The PFD is a dimensionless number between zero and one. As the value of the PFD decreases, there is a larger reduction in the likelihood that the incident scenario will continue to propagate and result in a consequence of concern. The reduction in the likelihood of scenario propagation may also be expressed as a "risk reduction factor" (RRF). A RRF is calculated as 1/PFD, which converts the fractional number into a whole number. (For example, a PFD of 0.1 is equivalent to a RRF of 10.)

The effectiveness and independence of an IPL can be validated and audited to ensure that the assumptions made in LOPA are demonstrated in the operating environment. Safeguards that do not meet the requirements to be counted as IPLs in LOPA can still be important to the organization as part of the overall risk reduction scheme and good engineering practice.

5.1.1 General Requirements for IPLs

Rules for IPLs: As indicated in *Guidelines for Safety and Reliable Instrumented Protective Systems* (CCPS 2007b) and discussed in Chapter 3, an IPL

- Is independent of other IPLs and the IE
- Functions in a way that prevents or mitigates the consequence of concern
- Has sufficient integrity to be capable of completely preventing the consequence of the scenario

- Can be relied upon to operate as intended, under stated conditions, for a specified time period
- Can be audited to ensure that the management systems to support the IPL are in place and effective
- Is protected by access security, with controls in place to reduce the chance of impairment
- Is covered by a management of change process to review, approve, and document changes

These core attributes for an IPL have been instrumental in establishing the LOPA methodology.

5.1.2   IPLs versus Safeguards

> All IPLs are safeguards, but not all safeguards are IPLs.

The distinction between an IPL and a safeguard is important. A safeguard is any device, system, or action that will likely either interrupt the chain of events following an initiating event or mitigate the consequences. The effectiveness of some safeguards cannot be quantified due to lack of data, uncertainty as to independence or effectiveness, or other factors. A safeguard is not an IPL if it does not meet the core attributes as described in Chapter 3.

Safeguards *should* be included in the scenario description as they may be candidates for being upgraded, making them into IPLs if the initial LOPA result does not totally meet the target risk goal. Further, safeguards might be called upon to serve as protection measures temporarily while IPLs are impaired due to maintenance activities.

As a final comment on "safeguards versus IPLs," it is best practice to always look for diverse types of IPLs and safeguards. Even if an IPL can be proven to be independent in all typical respects, there are common cause failures that may be difficult to predict or to detect if IPLs and safeguards are of the same type. For example, all instruments might be tested at the same time by a technician who uses the same miscalibrated equipment or consistently makes the same calibration error. In such situations, the reliability of similar devices can be decreased.

5.1.3   Basic Assumptions for IPLs

As summarized in Chapter 2, for an IPL to be valid, the organization needs to maintain the underlying management systems that ensure the IPL will meet the assigned value (PFD). The basic systems that greatly affect the reliability (and hence the unreliability) of an IPL are the same as those that affect an IE. For details, refer to Chapter 2.

5.2   SPECIFIC INDEPENDENT PROTECTION LAYERS FOR USE IN LOPA

Provided in the rest of this chapter are brief descriptions of many types of candidate IPLs, along with general descriptions of each IPL category and examples that illustrate when an IPL may or may not be appropriate for a scenario. Of course, each candidate IPL is also expected to meet the requirements for an IPL before a PFD value can be assigned. Also mentioned are candidate IPLs for which the *Guidelines* subcommittee could not offer a generic value. To consider such safeguards as potential IPLs, the LOPA analyst or user could use other quantitative risk assessment methods to assign the appropriate PFDs to such IPLs. (Refer to Chapter 6 for a discussion of these methods.)

Throughout this chapter are IPL data tables for candidate IPLs. Each IPL data table gives a brief description of each candidate IPL, provides a typical PFD for those IPLs, lists the criteria that are associated with each IPL value, and indicates a generic validation approach that can be used to maintain the reliability of the IPL at the claimed PFD. These PFDs are generally considered to be conservative values under the conditions specified. If, however, a site has data indicating that the actual IPL PFD is higher than that shown in the data table, the actual site value should be used. It also is possible that site-specific data collection or the use of a more quantitative risk assessment may indicate that a lower PFD may be used. Please refer to Appendices B and C for additional guidance on the development of site-specific equipment and human performance data. Each site should ensure that the IEF values selected are applicable to their facility.

In some cases, the PFDs used in the data tables were developed based upon published data; in this case, the reference for the data is provided. In other cases, the generic PFD and the guidance to maintain that value are based on research from within one or more companies represented on this *Guidelines* subcommittee. In all cases, the *Guidelines* subcommittee reached consensus on the generic values and guidance provided in each data table.

Companies, including some that participated in the writing of this document, use IPLs that are not listed in Chapter 5. Examples of some of these can be found in Chapter 6. See Section 5.3 for the general situation of "What If Your Candidate IPL Is Not Shown in a Data Table?"

CHAPTER 5 DATA TABLES

5.2.1   Passive IPLs

A passive IPL is not required to take an action in order for it to achieve its function in reducing risk. Such IPLs achieve the intended function if their process or mechanical design is correct and if constructed, installed, and maintained correctly. Examples are tank dikes, blast walls or bunkers, and flame or detonation arresters.

These devices are intended to prevent undesired consequences such as widespread leakage, blast or fire damage, or a flame front propagating to connected piping. If designed adequately, such passive systems can be credited as IPLs and will significantly reduce the frequency of events with potentially major consequences.

In general, the use of a passive IPL is only applicable if the consequence is defined in the IPL's absence (i.e., if the consequence severity can only result from the failure of the IPL to be effective). If the defined consequence already accounts for the passive feature, then the passive IPL cannot be credited in the LOPA. For example, if the consequence is stated as "flammable liquid spills into a diked area, leading to fire in the dike," then the consequence definition has already assumed that the dike is in place and effective in containing the spill. In this case, the dike would not be credited as an IPL in the scenario. For an alternate scenario, where the defined consequence is environmental contamination if a spill is not contained, then it may be appropriate to take credit for the dike as an IPL.

If a passive IPL is successful in preventing a specific consequence of concern, there still may be other potential consequences that will result from the successful use. These potential consequences should also be analyzed. Using the example above, the dike may have successfully prevented environmental contamination; however, the potential for a pool fire inside the dike or a vapor cloud resulting from the release would also need to be assessed.

Some passive IPLs, such as flame arresters, employ simple physical principles but are susceptible to fouling, plugging, corrosion, and unexpected environmental conditions. These failure modes should be considered when assigning PFDs to such devices.

Other passive IPLs, where the equipment design prevents the consequence (such as dikes and blast walls), can have low PFD values (high reliability) for LOPA purposes. However, care should be taken to assess the specific application to accurately determine the appropriate PFD to apply.

Static Dry Flame Arresters

A static dry flame arrester is an in-line passive protection device that allows vapor flow, but prevents the passage of flame. It functions by forcing a flame front through narrow channels; this slows and cools the flame front, stopping the propagation of the flame. It is important to select the appropriate type of flame arrester for the specific consequence being avoided. Categories include:

- *Deflagration:* A deflagration occurs when flammable vapors are ignited inside equipment or piping. Burning vapor heats and ignites the next segment of fuel, creating a flame front that propagates through equipment/piping at a subsonic velocity. Deflagration arresters can be used in-line or at the end of a pipe.

- *Stable Detonation:* A detonation occurs when the flame front accelerates from subsonic velocity to supersonic velocity, creating a shock wave. The point at which a flame front transitions from a deflagration to a detonation is known as the deflagration-to-detonation transition (DDT). The detonation is considered to be stable when the flame front passes through a zone without significant variation in velocity or pressure.

- *Unstable Detonation:* Immediately following DDT, there is a period where the pressure of the flame front is greater than that of the corresponding stable detonation flame front. This is generally a brief, temporary condition that is known as unstable (or overdriven) detonation. Conditions during this transition are more severe due to the higher shock wave pressure. Under some conditions, the flame front may be subject to repeated transitions between stable and unstable detonations; this is referred to as a "galloping detonation." Flame arresters designed for stable detonation may not be effective for unstable detonations.

The subsequent data tables cover four flame arrester applications:

Data Table 5.1. End-of-line deflagration arrester
Data Table 5.2. In-line deflagration arrester
Data Table 5.3. In-line stable detonation arrester
Data Table 5.4. Unstable (overdriven) detonation arrester

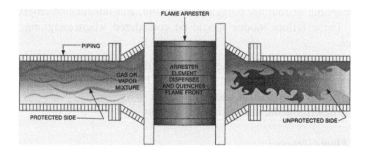

Figure 5.1. Flame arrester operation, courtesy of the Protectoseal Company.

End-of-Line Deflagration Arrester

Description: An end-of-line deflagration arrester is installed at the end of a pipe or an atmospheric vent discharge. It is typically placed between a flammable or combustible vapor inside equipment and the atmosphere. It allows for venting of the flammable or combustible material while preventing the passage of an external flame into the protected equipment. These devices can be used to protect against lightning or other transient ignition sources.

To ensure effective performance, the deflagration arrester needs to be tested and certified according to the protocols and specifications of a recognized organization. If used under conditions different from those for which it was listed or approved by a recognized organization, the device should be validated to ensure that the desired performance is achieved at the alternative conditions.

An end-of-line deflagration arrester is exposed to the environment, and it is important that it be kept clear of snow or ice. Devices should be periodically inspected for nests, hives, debris, polymer or other plugging material, and corrosion of the flame arrester element that could impair performance.

Consequences Avoided: Properly designed, installed, and maintained, an end-of-line deflagration arrester will prevent the passage of an external flame into the protected equipment and prevent a deflagration within the equipment. Deflagration arresters will not protect against detonations. Also, ignition of flammable vapors outside of the protected equipment is not prevented by a deflagration arrester.

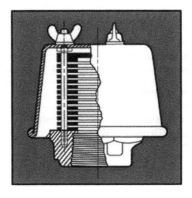

Figure 5.2. End-of-line deflagration arrester, courtesy of the Protectoseal Company.

Data Table 5.1. End-of-line deflagration arrester

IPL description
Deflagration arrester at the end of pipe, typically between the location of the ignitable vapors and potential ignition sources.
Generic PFD suggested for use in LOPA
0.01
Special considerations for use of generic PFD for this IPL
• The deflagration arrester is installed in the appropriate location and orientation as per the manufacturer's recommendations. • The device does not impose excessive flow restriction on the process, and any fouling issues have been addressed. • The device is used in the service for which it has been certified and tested. • An end-of-line device needs to be kept clear of ice and snow, debris, and nesting creatures that could block the device and impair its performance. • Elements are periodically inspected for polymer or other plugging material as well as for internal corrosion of the flame arrester element.
Generic validation method
• ITPM frequency is set in accordance with the manufacturer's recommendations and based on the results of previous inspections. • The device is included on a routine maintenance and inspection schedule. The inspection includes determining whether the device is plugged and whether corrosion might compromise its capability to arrest a deflagration. • The device is immediately inspected if it is suspected to have stopped a flame, or if a process upset could have compromised its integrity.
Basis for PFD and generic validation method
Consensus of the *Guidelines* subcommittee. Reference *Deflagration and Detonation Flame Arresters* (Grossel 2002).

In-Line Deflagration Arrester

Description: The arrester is installed in the piping between a potential ignition source and a source of flammable or combustible vapors. A deflagration arrester can be used when the distance between the flammable vapor source and the arrester (called the "run-up distance") is too short to allow the flame front to accelerate and transition to a detonation.

To ensure effective performance, the deflagration arrester has been tested and certified according to the protocols and specifications of a recognized organization. If used under conditions different from those for which it was listed or approved by a recognized organization, the use of the device is validated to ensure that the desired performance is achieved at the alternative conditions. For proper operation, the device is installed as per the vendor recommendations including the location in the line and orientation (horizontal or vertical).

Deflagration arresters cannot withstand flame indefinitely, and arrester elements generally need to be replaced after they are challenged since they may have been damaged and may not be effective in stopping a second challenge. The reliability of a device can be improved by implementing a means to detect that a device is being challenged and a response that prevents continued flame impingement. Temperature detection will also facilitate maintenance so that the device can be return to an "as good as new" state after the event. Data Table 5.2 suggests a generic PFD value of 0.01 for an in-line deflagration arrester with temperature indication and an effective shutdown or isolation response. When temperature indication and an effective response is not available, a generic PFD value of 0.1 is suggested.

Consequences Avoided: Properly designed, installed, and maintained, an in-line deflagration arrester will prevent the passage of a subsonic flame front from associated piping into the protected equipment, preventing a deflagration within the protected equipment. Deflagration arresters will not protect against detonations.

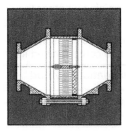

Figure 5.3. In-line deflagration arrester, courtesy of the Protectoseal Company.

Data Table 5.2. In-line deflagration arrester

IPL description
Deflagration arrester in-line between a potential ignition source and equipment containing flammable or combustible vapors.
Generic PFD suggested for use in LOPA
0.1 without temperature monitoring and shutdown or isolation response 0.01 with temperature monitoring and shutdown or isolation response
Special considerations for use of generic PFD for this IPL
• The piping between a potential ignition source and the arrester is well below the run-up distance required to allow a transition to detonation (DDT) and does not include turbulence-inducing fittings that could cause DDT. • The deflagration arrester is installed in the appropriate location and orientation as per the manufacturer's recommendations. • The device does not impose excessive flow restriction on the process, and any fouling issues have been addressed. • Temperature monitoring with a thermocouple directly in contact with the hot side of the device is recommended to allow operations to recognize when the device is being challenged. Use of temperature monitoring with response increases the reliability of the device. • The device is used in chemical services for which it has been certified and tested. Detonation arresters built to meet older standards (pre-1990s) are not guaranteed to prevent passage of a deflagration.
Generic validation method
• The ITPM frequency is set in accordance with the manufacturer's recommendations and based on the results of previous inspections. • The device is included on a routine maintenance schedule that specifies opening the device for inspection. • The inspection includes determining whether the device is plugged, and whether corrosion might compromise its capability to arrest a deflagration. • The device is immediately inspected if it is suspected to have stopped a flame, or if a process upset could have compromised its integrity.
Basis for PFD and generic validation method
Consensus of the *Guidelines* subcommittee. Reference *Deflagration and Detonation Flame Arresters* (Grossel 2002).

In-Line Stable Detonation Arrester

Description: The in-line stable detonation arrester is installed in piping between a potential ignition source and a source of flammable or combustible vapors. It is used when the piping and equipment configuration is such that a deflagration-to-detonation transition (DDT) of a flame front cannot be ruled out. For example, a vent system header directing potentially flammable vapors to a thermal oxidizer could be protected by an in-line detonation arrester.

To ensure effective performance, the detonation arrester has been tested and certified according to the protocols and specifications of a recognized organization. If used under conditions different from those for which it was listed or approved by a recognized organization, the use of the device should be validated to ensure that the desired performance is achieved at the alternative conditions. For proper operation, the device is installed as per the vendor recommendations including the location in the line and orientation (horizontal or vertical).

Deflagration arresters are not suitable for applications where a stable detonation can occur. Prior to crediting an arrester as an IPL for a scenario where a stable detonation can occur, there should be confirmation that the arrester is designed for stable detonation.

Detonation arresters cannot withstand flame indefinitely, and arrester elements generally need to be replaced after they are challenged since they may have been damaged and may not be effective in stopping a second challenge. The reliability of a device can be improved by implementing a means to detect that a device is being challenged and a response that prevents continued flame impingement. Temperature detection will also facilitate maintenance so that the device can be returned to an "as good as new" state after the event. Data Table 5.3 suggests a generic PFD value of 0.01 for an in-line detonation arrester with temperature indication and an effective shutdown or isolation response. When temperature indication and an effective response is not available, a generic PFD value of 0.1 is suggested.

Consequences Avoided: Properly designed, installed, and maintained, an in-line stable detonation arrester will prevent the passage of a supersonic flame front from associated piping into the protected equipment, preventing a detonation within the equipment.

Data Table 5.3. In-line stable detonation arrester

IPL description
In-line stable detonation arrester between a potential ignition source and equipment containing flammable or combustible vapors, where deflagration-to-detonation transition (DDT) cannot be ruled out.
Generic PFD suggested for use in LOPA
0.1 without temperature monitoring and shutdown or isolation response 0.01 with temperature monitoring and shutdown or isolation response
Special considerations for use of generic PFD for this IPL
• The device is installed in the appropriate location and orientation as per the manufacturer's recommendations. • The device does not impose excessive flow restriction on the process, and any fouling issues have been addressed. • Temperature monitoring with a thermocouple directly in contact with the hot side of the device is recommended to allow operations to recognize when the device is being challenged. Use of temperature monitoring with response increases the reliability of the device. • The device is used in chemical services for which it has been certified and tested.
Generic validation method
• The ITPM frequency is set in accordance with the manufacturer's recommendations and based on the results of previous inspections. • The device is included on a routine maintenance schedule that specifies opening the device for inspection. • The device is immediately inspected if it is suspected to have stopped a detonation, or if a process upset could have compromised its integrity. • The inspection includes determining whether the device is plugged, and whether corrosion might compromise its capability to arrest a detonation.
Basis for PFD and generic validation method
Consensus of the *Guidelines* subcommittee. Reference *Deflagration and Detonation Flame Arresters* (Grossel 2002).

Unstable (Overdriven) Detonation Arrester

Description: An unstable (overdriven) detonation arrester is installed in-line between an ignition source (e.g., a thermal oxidizer) and a source of flammable or combustible vapors. They are used when a DDT could occur and where an unstable detonation condition is possible within the arrester.

To ensure effective performance, the unstable detonation arrester is tested and certified prior to installation according to the protocols and specifications of a recognized organization. If used under conditions different from those for which it was specified, the use of the device at the alternate condition is validated. The device is installed as per the vendor recommendations, including location in the line and orientation (horizontal or vertical).

Unstable detonation arresters are able to stop deflagrations, stable detonations, and unstable detonations. Detonations pass through an unstable regime before reaching a stable zone. Stable detonations can also transition through an unstable regime in regions of high turbulence. Unstable detonation arresters provide protection if a detonation has not yet become stable prior to reaching the arrester, or if the detonation returns to an unstable regime. If properly designed for the specific application by a qualified vendor and subject to the rigorous testing required of an unstable (overdriven) detonation arrester, the PFD for these devices is one order of magnitude lower than the PFD for other arresters.

Deflagration arresters are not suitable for applications where a detonation can occur. Also, some detonation arresters are designed for stable detonations but are not suitable for protection against unstable (overdriven) detonation. Prior to crediting an arrester as an IPL for a scenario where an unstable detonation can occur, there should be confirmation that the arrester is designed for an unstable detonation for the specific application.

Detonation arresters cannot withstand flame indefinitely, and arrester elements generally need to be replaced after they are challenged since they may have been damaged and may not be effective in stopping a second challenge. The reliability of a device can be improved by implementing a means to detect that a device is being challenged and a response that prevents continued flame impingement. Temperature detection will also facilitate maintenance so that the device can be return to an "as good as new" state after the event. Data Table 5.4 suggests a generic PFD value of 0.001 for an in-line unstable detonation arrester with temperature indication and an effective shutdown or isolation response. When temperature indication and an effective response is not available, a generic PFD value of 0.01 is suggested.

Consequences Avoided: An in-line unstable detonation arrester can prevent the passage of an unstable, supersonic flame front from associated piping into the protected equipment, preventing a detonation within the equipment.

Data Table 5.4. Unstable (overdriven) detonation arrester

IPL description
Unstable (overdriven) detonation arrester installed in-line between an ignition source and a source of flammable or combustible vapors.
Generic PFD suggested for use in LOPA
If properly designed for the specific application by a qualified vendor and subject to the rigorous testing required of an unstable (overdriven) detonation arrester, the PFD for these devices is one order of magnitude lower than the PFD for other arresters. The suggested PFDs are: 0.01 without temperature monitoring and shutdown or isolation response 0.001 with temperature monitoring and shutdown or isolation response
Special considerations for use of generic PFD for this IPL
• The device is installed as per the vendor recommendations, including location in the line and orientation (horizontal or vertical). • The device does not impose excessive flow restriction on the process, and any fouling issues have been addressed. • Temperature monitoring with a thermocouple directly in contact with the hot side of the device, with reliability commensurate with the claimed value for the system, is recommended to allow operations to recognize when the device is being challenged. Unstable detonation arresters require specific empirical determination of proper sizing for each specific stream composition.
Generic validation method
• The ITPM frequency is set in accordance with the manufacturer's recommendations and based on the results of previous inspections. • Input is required from the supplier to determine the maintenance activities and frequencies required to achieve and maintain the higher reliability required for an unstable detonation arrester. • The device is included on a routine maintenance schedule that specifies opening the device for inspection. • The device is immediately inspected if it is suspected to have stopped a detonation, or if a process upset could have compromised its integrity. Inspection includes determining whether the device is plugged, and whether corrosion might compromise its capability to arrest a detonation.
Basis for PFD and generic validation method
Consensus of the *Guidelines* subcommittee. Reference *Deflagration and Detonation Flame Arresters* (Grossel 2002).

Overflow Lines

Overflow lines are passive IPLs that release material from a tank in the event of overfilling. Some overflow lines are simple, open pipes that drain material from the tank. However, the open pipe can provide a path for vapors to be discharged from the tank and may be best suited for nonhazardous material service. To prevent vapors from being discharged from the tank, some overflow lines contain a loop that is filled with a passive fluid, such as mineral oil. The fluid allows for liquid overflow while preventing emissions. The reliability of such systems is best when the seal fluid is not prone to freezing or evaporation, and the vapors generated inside the tank are not prone to fouling or polymer build-up. A vapor barrier can also be created in an overflow line by the use of a rupture disk.

Overflow lines can be effective IPLs to prevent the consequence of overfilling a tank. However, the successful operation of an overflow may create a secondary consequence as a result of a liquid release from the tank. It may be appropriate to add this scenario to the LOPA as well.

Three types of overflow systems are covered by the following data tables:

Data Table 5.5. Overflow line with no impediment to flow

Data Table 5.6. Overflow line containing a passive fluid or with a rupture disk

Data Table 5.7. Line containing a fluid with the potential to freeze

Description: A simple overflow line with no impediment to flow. There are no valves, rupture disks, or seal legs that could cause a restriction of flow. The overflow line is sized for the maximum storage tank fill rate.

Consequences Avoided: A properly sized overflow line can prevent overpressurization of a tank or vessel due to overfilling.

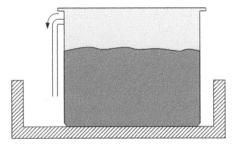

Figure 5.4. Illustration of an overflow line with no impediment to flow.

Data Table 5.5. Overflow line with no impediment to flow

IPL description
Overflow line with no impediment to flow
Generic PFD suggested for use in LOPA
0.001
Special considerations for use of generic PFD for this IPL
The overflow line is properly sized with no valves that could be closed, seal legs that could be frozen or blocked, or rupture disks that could fail to operate. • This failure rate assumes that the overflow line is in a clean service and not prone to fouling, plugging, or polymerization. • The overflow line is designed to allow for inspection and to prevent blockage due to snow, ice, nests, or other debris. • The overflow line does not have a vertical rise above the top of the vessel or drum that could result in the sum of the hydraulic and dynamic pressure being greater than the MAWP of the vessel.
Generic validation method
• The inspection frequency is determined based on previous inspection history and environmental conditions. • Inspections are done to ensure that there is no plugging or fouling of the overflow line and that no impediments to flow have been added. • Inspection is performed to verify there is no blockage in the line.
Basis for PFD and generic validation method
Consensus of the *Guidelines* subcommittee. See *Guidelines for Safe and Reliable Instrumented Protective Systems* (CCPS 2007b), p 288, Table B.4.

Vessel Overflow Line Containing a Passive Fluid or with a Rupture Disk

Description: This type of overflow line contains a passive, nonvolatile, clean fluid, such as mineral oil, with no valves en route to the discharge location. The seal fluid is not subject to freezing or fouling. The seal fluid acts as a barrier, to prevent vapors from exiting the tank and entering the environment. An overflow line with a rupture disk (RD) installed to act as a vapor barrier is also covered by this IPL data table.

Consequences Avoided: A properly sized overflow line can prevent overpressurization of a tank or vessel due to overfilling.

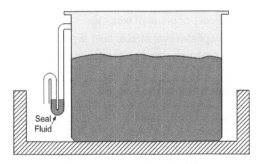

Figure 5.5. Illustration of an overflow line with a seal fluid.

Data Table 5.6. Overflow line containing a passive fluid or with a rupture disk

IPL description
Overflow line containing a passive fluid or with a rupture disk

Generic PFD suggested for use in LOPA
0.01

Special considerations for use of generic PFD for this IPL
The overflow line is properly sized with no valves that could be closed or seal legs with fluid that could freeze or foul.

- This failure rate assumes that the overflow line is in a clean service and not prone to fouling, plugging, or polymerization.
- The overflow line is designed to allow for inspection and to prevent blockage due to snow, ice, nests, or other debris.
- The overflow line does not have a vertical rise above the top of the vessel or drum that will result in the sum of the hydraulic and dynamic pressure being greater than the MAWP of the vessel.
- For installations including rupture disks, the criteria and considerations provided in the data tables for rupture disks also apply. The rupture disk opening pressure needs to be low enough that the vessel will not be overpressurized if overfilling does actually occur.

Generic validation method

- The inspection frequency is set based on the manufacturer's recommendation and prior inspection results.
- Inspection shows no history or onset of blockage.
- For a seal leg, the appropriate level of seal fluid is present.
- For a rupture disk, the pressure rating on the tag is checked and the rupture disk is intact.

Basis for PFD and generic validation method
Consensus of the *Guidelines* subcommittee. See *Guidelines for Safe and Reliable Instrumented Protective Systems* (CCPS 2007b), p 288, Table B.4. The PFD is an order of magnitude higher than the PFD for an overflow line with no impediment due to the potential for blockage in the line.

Overflow Line Containing a Fluid with the Potential to Freeze

Description: The design of this type of device includes overflow piping from a tank/vessel/drum with additional hardware to form a liquid seal leg. This data table covers an overflow line that contains a fluid that could, if not properly managed, freeze and create a blockage in the overflow line. There may be means (such as heat tracing) to prevent freezing. This IPL may also include one or more valves in the overflow path that could act as an impediment to flow.

Where a system is prone to fouling, a generic IPL PFD value cannot be provided. The performance of the system is highly dependent on the specific service, the characteristics of the fluid, the tendency of the piping to foul, and the programs in place to ensure ongoing integrity and to confirm that any isolation valves are maintained in the open position during operation.

Consequences Avoided: A properly sized overflow line can prevent overpressurization of a tank or vessel due to overfilling.

Data Table 5.7. Line containing a fluid with the potential to freeze

IPL description
Line containing a fluid with the potential to freeze
Generic PFD suggested for use in LOPA
0.1
Special considerations for use of generic PFD for this IPL
The overflow line is properly sized, and there is a possible impediment to flow such as a seal leg that could freeze or a valve that could be closed. • It is assumed that reasonable design and maintenance precautions are taken to prevent situations that could result in freezing or an improperly closed valve. • The system is designed to allow for periodic inspection. • The PFD is higher than the PFD of the simple overflow line due to the increased failure rate associated with possible isolation of the line as a result of human error, mechanical failure, or weather. • The overflow line does not have a vertical rise above the top of the vessel or drum that will result in the sum of the hydraulic and dynamic pressure being greater than the MAWP of the vessel.
Generic validation method
• The inspection frequency is set based on the manufacturer's recommendation and prior inspection results. • Inspections are done to detect plugging or fouling of the overflow line. • Inspections are done on the valves and seal legs on the overflow line to ensure proper operation. • The appropriate level of seal fluid is present. • Inspection during cold weather is done to ensure that the method or procedure to prevent freezing is being followed and that freezing has not occurred.
Basis for PFD and generic validation method
Consensus of the *Guidelines* subcommittee. See *Guidelines for Safe and Reliable Instrumented Protective Systems* (CCPS 2007b), p 288, Table B.4.

Dikes, Berms, and Bunds

Description: Dikes, berms, and bunds are containment systems, usually consisting of a retaining wall and base/floor, that are capable of containing a spill of material and preventing its spread to a wider area. This can reduce the evaporation rate of the material, reduce the chance of ignition, and limit impact to equipment, people, and the environment. A containment system is a candidate for an IPL for LOPA if it prevents the consequence of concern, and a PFD of 0.01 may be appropriate if all special considerations are met. To claim a dike, berm, or bund as an IPL, the containment system must fail for the final consequence of the scenario to be reached. If the consequences are fully achievable without the containment system failing (such as loss of containment from a vessel), then the system cannot be given credit as an IPL.

For the case of vessel failure, containment systems are generally sized to contain the greatest amount of liquid that may be released from the largest tank within the diked area, providing sufficient allowance for precipitation (refer to NFPA 30 [NFPA 2008a]). The requirement for freeboard capacity is often set by local jurisdictions. For a catastrophic failure of a vessel, the containment system height should be able to withstand the hydraulic wave effects, with minimal slosh over the dike walls. If drain valves exist, they can be car sealed or locked in the correct position to ensure that they remain closed.

Consequences Avoided: Liquid material that is released into the containment is prevented from flowing outside of the designated area. The containment needs to be effective in preventing the specific consequence of the scenario. For instance, a containment system may prevent a significant process safety consequence for a liquid released below its boiling point, but for compressed gases or liquids stored above their atmospheric boiling point, the system may not be an effective IPL.

Figure 5.6. Illustration of a storage tank dike (U.S. EPA 2009).

Data Table 5.8. Dikes, berms, and bunds

IPL description
Dikes, berms, and bunds
Generic PFD suggested for use in LOPA
0.01
Special considerations for use of generic PFD for this IPL
A management system is in place to ensure that any drain valves on passive containment are maintained in the correct position and are subject to a valve inspection program. • For the case of vessel failure, the containment systems are typically sized to contain the greatest amount of liquid that may be released from the largest storage tank within the diked area, providing sufficient allowance for precipitation (refer to NFPA 30 [NFPA 2008a]). • To use the containment system as an IPL in an overflow scenario in which the source of the overflow has a greater volume than the maximum dike capacity of the receiving tank (such as a pipeline), the capacity of the system is sufficient to ensure that the leak is detected before the dike capacity is exceeded. • The containment system height is able to withstand the hydraulic wave effects, with minimal slosh over the dike walls.
Generic validation method
• Visual inspection is needed to confirm the integrity of the containment system and ensure proper valve position. • Mechanical/civil inspections of the containment or barrier are conducted at an appropriate frequency. • Periodic management system reviews can ensure the effectiveness of the dike valve management program.
Basis for PFD and generic validation method
Consensus of the *Guidelines* subcommittee.

Drainage to Dikes, Berms, and Bunds with Remote Impoundment

While a dike, berm, or bund is usually capable of containing hazardous liquid spills, it can also serve to pool flammable liquid fuel underneath tanks. If ignited, the resulting pool fire can overheat the tanks and cause them to fail. Therefore, dikes, berms, and bunds are generally ineffective IPLs for pool fire scenarios and for boiling liquid expanding vapor explosion (BLEVE) scenarios for liquefied flammable gases. Dikes, berms, and bunds are also generally shallow, requiring large surface areas to hold the necessary volumes. The combination of wind with a large liquid surface area can result in significant evaporation rates and possible downwind effects (ignition and/or toxic impact). Remote impoundment is designed to address these specific hazards by directing the flow of the hazardous liquid to a deep pit with minimal surface area. The pit is generally located away from the vessel(s) being protected. The walls sometimes extend above ground height to minimize wind velocity on the pool surface and to minimize or eliminate radiation impact to surrounding equipment should the pool ignite. A smaller surface area allows for extinguishment of the fire with minimal use of firefighting foam.

Remote impoundment is typically a candidate for an IPL for LOPA in lieu of a dike, berm, or bund and can have a PFD of 0.01 if all of the special considerations are met. A critical aspect of this IPL is ensuring that the flow path to remote impoundment remains clear and open, with no closed drain valves or other diversion to impede the flow of liquid. Often, grating is placed over the receiving trench or sump within the dike, and periodic maintenance can be effective in keeping grating from becoming blocked with leaves and other debris.

A remote impoundment basin should be sized so that it can contain the greatest amount of liquid that may be released from the largest tank within the diked area (assuming a full tank) and also provide sufficient allowance for precipitation. The requirement for freeboard capacity is often set by local jurisdictions or by industry standards such as NFPA 30 (2008a). For scenarios in which the source of the overflow can potentially supply a greater volume than the maximum dike capacity of the receiving tank (such as a pipeline), the capacity of the containment system is sufficient to ensure that the leak is detected before the dike capacity is exceeded. The dike is constructed such that it will contain the liquid that may be spilled and channel it to the sump or trench, where it can flow by gravity to the remote impoundment. In the event of a catastrophic failure of a vessel, the dike height can withstand the hydraulic wave effects with minimal slosh over the dike walls. Dikes are constructed with dike seals and expansion joints that are chemically resistant to contained chemicals.

Consequences Avoided: Any liquid that is released into the containment is prevented from flowing outside of the designated containment and from pooling and igniting underneath equipment. Remote containment can also reduce the potential surface area of a spill by providing a means to minimize vaporization of volatile liquid spills and subsequent downwind effects.

Data Table 5.9. Drainage to dikes, berms, and bunds with remote impoundment

IPL description
Drainage to dikes, berms, and bunds with remote impoundment
Generic PFD suggested for use in LOPA
0.01
Special considerations for use of generic PFD for this IPL
A management system is in place to ensure that any drain valves on a passive containment system to remote impoundment are subject to a valve inspection program and are maintained in the appropriate positions. • Rainwater is pumped from the deep impoundment to the appropriate sewer system to maintain adequate spill containment capacity. • Any pipe or trench used to transmit liquid to the remote impoundment is kept free of impediments to flow, such as insulation materials or debris from trees and shrubs. • For the case of vessel failure, the dike plus impoundment volume is generally sized to contain the greatest amount of liquid that may be released from the largest storage tank within the diked area, providing sufficient allowance for precipitation (refer to NFPA 30 [NFPA 2008a]). • To use the containment system as an IPL in an overflow scenario in which the source of the overflow has a greater volume than the maximum dike capacity of the receiving tank (such as a pipeline), the capacity of the system is sufficient to ensure that the leak is detected before the dike capacity is exceeded. • The containment system height is able to withstand the hydraulic wave effects, with minimal slosh over the dike walls.
Generic validation method
• Visual inspection of the containment system and transmission pipes, trenches, and grating is done, with immediate removal of debris. • Mechanical/civil inspection of the containment or barrier may be periodically required.
Basis for PFD and generic validation method
Consensus of the *Guidelines* subcommittee.

Permanent Mechanical Stop That Limits Travel

Description: A permanent mechanical stop is a device that limits travel of a component, such as a piston, valve, or machine. For example, in the event that complete closure of a valve would result in an undesirable consequence (such as in applications using heater tubes in furnaces and in some reactor loop systems), a stop can be welded inside the valve to ensure that the valve cannot be completely shut. The mechanical stop is permanently secured and cannot be removed, relocated, or reset. Mechanical stops represent a diverse layer of protection that is not related to human or instrumented IPLs.

Some mechanical stops are adjustable; these are less reliable since they can be more easily removed or reset by workers. Adjustable movement limiting devices are discussed in Data Table 5.48.

Consequences Avoided: The permanent mechanical stop prevents movement or travel beyond specified limit, preventing a consequence of concern.

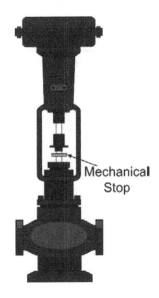

Figure 5.7. Mechanical stop.

Data Table 5.10. Permanent mechanical stop that limits travel

IPL description
Permanent mechanical stop that limits travel
Generic PFD suggested for use in LOPA
0.01
Special considerations for use of generic PFD for this IPL
• The stop is set and verified initially, with adjustment or slippage not possible afterwards (such as welding the stop in place). • It is important that the reason for the placement of the mechanical stop is understood and documented so that the stop is not inadvertently eliminated when equipment is replaced.
Generic validation method
• The actual device placement is measured periodically to ensure that the device remains in the appropriate location. • Components are inspected to ensure ongoing integrity. • Visual examination is performed to detect erosion/corrosion/wear and to validate the stop placement.
Basis for PFD and generic validation method
Consensus of the *Guidelines* subcommittee.

Fire-Resistant Insulation and Cladding on Vessel

Fireproofing of tanks, columns, and structures decreases the heat transferred to the protected equipment or structure and therefore increases the time before a heat-induced failure occurs. Insulation is installed in such a manner that firefighting water will not dislodge it from the equipment or result in a loss of insulation value due to water soaking.

If the fireproofing can last long enough to allow consumption of the fuel (so that the fire goes out prior to the consequence of concern), then fireproofing can be credited as an IPL for LOPA. If the fuel is not consumed, then slowing down heat-induced failure may allow other IPLs to be valid, even if fireproofing is not itself credited as an IPL. An example of this may occur if fireproofing provides sufficient time for fire suppression systems or human response IPLs to be effective.

Description: Fire-resistant insulation and cladding are means to reduce the rate of heat input to equipment. They may be useful when considering the appropriate sizing basis for relief valves, preventing a boiling liquid expanding vapor explosion (BLEVE), or preventing an exothermic runaway reaction due to external heat input. If the amount of fuel available for an external fire is known, the fire duration can be estimated. If the fire-resistant insulation reduces the heat flux from a known quantity of fuel to a value that will not result in overpressure of the vessel or heat the vessel contents to a temperature that may initiate a runaway reaction, fire-resistant insulation can be an effective IPL for LOPA. Refer to ANSI/API 521 (2008) for such calculations.

Consequences Avoided: Proper installation and maintenance of fireproofing can provide sufficient time for the external fuel to be consumed, preventing vessel overpressure or a runaway reaction.

NOTE: If the size of a relief device used in a LOPA scenario has been decreased because of the assumed presence of fire-resistant insulation and cladding, then the fire-resistant insulation and cladding would not be considered to be a separate IPL, independent of the relief device. Rather, it would be part of the relief device IPL.

Data Table 5.11. Fire-resistant insulation and cladding on vessel

IE description
Fire-resistant insulation and cladding on vessel
Generic IEF suggested for use in LOPA
0.01 NOTE: If fire cladding was assumed in the calculation of the size of a relief valve that is claimed as an IPL in the LOPA scenario, then the combined PFD of the insulation plus relief valve is 0.01 (without a block valve in the relief line).
Special considerations for use of generic IEF for this IE
ANSI/API 521 (2008) allows the use of fire-resistant insulation as an alternative to a relief device sized for the fire case when an engineering analysis indicates that additional protection provided by the relief device serves little value in reducing the likelihood of vessel rupture. Vessels containing only vapor or high-boiling liquids are mentioned as specific examples. • In some instances, only a limited amount of fuel is available to sustain a fire for a given period of time (the *fire duration*). The fire duration can be determined from a conservative estimate of the available fuel, combined with fuel burning rates available in the literature (e.g., Mudan 1984). • ANSI/API 521 (2008) also provides equations for calculating the heat input to the vessel, based on insulation conductivity and thickness, that can be used to calculate the time required to heat the vessel contents to a vapor pressure corresponding to the vessel MAWP or relief device setpoint. When this heat-up time (with the fire insulation installed) is greater than the length of the fire, based on burning rate and the amount of fuel available, fireproof insulation is a potential IPL to prevent overpressure and rupture of the vessel due to fire. Process or storage vessels may contain materials that could, at elevated temperatures, react and generate heat or rapidly decompose. • This could result in potentially generating pressures or temperatures that may exceed the vessel MAWP and may not be relievable with conventional pressure relief devices. • Appropriate thermal stability testing may be necessary to establish conservative safe temperature limits. • If fireproofing would prevent the vessel from reaching an unsafe temperature before the fuel available was consumed, then it could be a valid IPL in LOPA.
Generic validation method
Routine visual inspection and periodic mechanical/civil inspection are performed.
Starting source of guidance
Consensus of the *Guidelines* subcommittee. Also refer to *CCPS LOPA* (2001).

Double-Walled Systems

Description: Double-walled systems can be found on piping and on vessels. This containment system includes a complete, secondary enclosure around the primary containment that reduces the likelihood of a complete loss of containment if the first containment wall fails. Although such a secondary containment system can be quite effective, it is not normally a stand-alone IPL for LOPA. Double-walled systems depend greatly on the method for detecting leakage into the interstitial space and the procedure for responding to a leak. The double wall would be considered to be part of an IPL that would also contain leak detection and response. For such an IPL, refer to the data table that is appropriate for the type of alarm or interlock being used in conjunction with the double-walled system.

Containment Buildings

Description: Like double-walled construction, the evaluation of containment buildings is highly specific to the design of the structure, the control of entrances, the controls of vents, and the contents of the building. Although a containment building can decrease the risk to those outside of the building, they can increase the risk to those who work inside of the containment building. For these reasons, there is no generic PFD value that can be provided for containment buildings. Each design needs to be individually evaluated. More quantitative risk assessment techniques may be required to ensure that the design and control can effectively manage the risk at a tolerable level. Refer to Chapter 6 for a discussion of such quantitative methods.

Blast-Resistant Building Designs

Description: These buildings have been designed to withstand a predicted overpressure to limit the consequence severity of an explosion. Each design needs to be individually evaluated to ensure that it will be effective against a specific scenario of concern. It may be necessary to conduct a consequence analysis, employing blast modeling to predict the maximum potential overpressure and impulse duration and comparing these data to existing building design criteria. A common risk assessment approach is to assume that all people inside a well-maintained, blast-resistant enclosure that is properly designed for the maximum anticipated blast loading are fully protected. Therefore, a LOPA scenario evaluating impact on those individuals due to a blast would not be necessary.

Blast (Sacrificial) Walls or Explosion Barriers

Description: These are barriers that have been designed to deflect or withstand a predicted overpressure and/or shrapnel impact, thereby limiting the consequence severity of a blast. Blast or explosion barriers are normally not used as IPLs in LOPA. Similar to blast-resistant building design, it may be appropriate to employ a blast modeling consequence analysis and compare the output to the building design criteria.

5.2.2   Active IPLs

Active IPLs ensure the functional safety of the process by taking action to prevent the scenario from occurring. This action may be mechanical-only in nature or may use a combination of instruments, humans, and mechanical devices.

5.2.2.1   Safety Controls, Alarms, and Interlocks (SCAI) as IPLs

LOPA involves the analysis of the sequence of events leading to a consequence of interest. This analysis yields an understanding of the role that automation plays in detecting and responding to each scenario. Consequently, LOPA facilitates the identification of instrumented systems as IEs and as instrumented safeguards. SCAI are process safety safeguards implemented with instrumentation and controls and used to achieve or maintain a safe state for a process. SCAI provides risk reduction with respect to a specific scenario of concern (ANSI/ISA 84.91.01-2012).

SCAI are a subset of the instrumented protective systems (IPS) that are implemented to address a wide variety of risks, including environmental, business, and process safety risks. As shown in Figure 5.8, there are many terms that can be used to further classify SCAI.

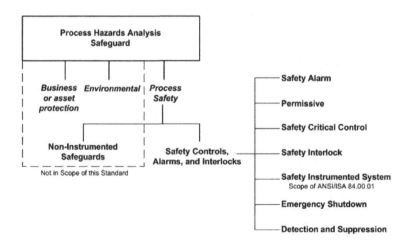

Figure 5.8. Safety controls, alarms, and interlocks relationship to the process hazard analysis (ANSI/ISA 84.91.01-2012).

Regardless of what they are called, SCAI are complex IPLs comprised of multiple devices that need to operate correctly when required. SCAI include sensors, logic solvers, final elements, and other interconnected equipment, such as human interfaces, wiring, process connections, and utilities. In the case of safety alarms, successful SCAI operation requires human intervention – an operator that acknowledges the process condition and takes prescribed actions, such as taking a manual control action or initiating a manual shutdown. The ability of SCAI to achieve the desired functionality and the claimed risk reduction is limited by the performance achieved by the installed equipment and the operator response, where applicable. For this reason, equipment is selected based on its expected capability in the operating environment, but the SCAI is judged by the cumulative capability of the system.

SCAI are most commonly implemented in electrical, electronic, or programmable electronic systems. In addition, SCAI may involve equipment that requires pneumatic, mechanical, and/or hydraulic systems to operate as required. SCAI use technology and equipment that have been demonstrated to provide the desired level of risk reduction in the operating environment.

Guidelines for Safe and Reliable Instrumented Protective Systems (CCPS 2007b) discusses the design and management requirements for SCAI in detail and describes the management systems needed to achieve target performance claims. This section provides a brief overview of these requirements and references specific clauses of IEC 61511 (2003), which limit risk reduction claims for instrumented protective systems that are not compliant with IEC 61511 (2003). These limits are due to the potential for systematic errors to impact the performance of the IPS (or SCAI) during its life. The functional safety management system of IEC 61511 (2003) goes beyond typical process safety management practices and requires rigorous assessment, verification, validation, and change management for any activity potentially impacting the SIS.

SCAI can be implemented using a variety of equipment, and the type of equipment and the design and management practices used to implement equipment affect their achievable performance. For example,

- BPCS: The primary intent of the equipment design and management is to keep the process within the normal operating range, e.g., PID (proportional-integral-derivative) control and sequential control. Failures of the BPCS in executing normal control functions are one of the leading causes of incidents. Refer to Data Table 4.1 for a discussion of BPCS control loop failure as an IE.

- SIS: The primary intent of the equipment design and management is to ensure the SIS can reliably take action when a demand occurs. In the process industrial sector, the term SIS applies to controls, alarms, and interlocks designed and managed in accordance with international standard IEC 61511 (2003).

This section provides limited guidance related to the equipment and practices, so the reader is referred to the following publications related to SCAI:

- *Guidelines for Safe and Reliable Instrumented Protective Systems* (CCPS 2007b)
- *Guidelines for Safe Automation of Chemical Processes* (CCPS 1993)
- ANSI/ISA 18.2 – *Management of Alarm Systems for the Process Industries* (ANSI/ISA 2009)
- ANSI/ISA 84.91.01 – *Identification and Mechanical Integrity of Safety Controls, Alarms, and Interlocks in the Process Industry* (ANSI/ISA 2012)
- ISA TR84.00.04 – *Guidelines for the Implementation of ANSI/ISA-84.00.01-2004 (IEC 61511 Mod)* (ISA 2011)
- IEC 61511 – *Functional Safety – Safety Instrumented Systems for the Process Industry Sector* (IEC 2003)

In evaluating a SCAI as a potential IPL, the team examines how the SCAI stops the scenario propagation (e.g., what condition does it detect, what is the setpoint, what action does it take on the process, how fast does it need to act, and how does the process respond to its action). As with any IPL, SCAI are independent of the IE and any other IPL used for risk reduction in the same scenario.

SCAI specifications consider the following:

- Functional requirements necessary to stop scenario progression
- Configuration, installation, and maintenance requirements to achieve and sustain the claimed performance
- Failure modes, means used to detect these failure modes, and expected system and operator response to detected failure
- Compensating measures required to continue safe operation when a fault exists in the SCAI
- Conditions required to safely bypass the SCAI (includes override or manual operation) and any compensating measures to be in place during bypass
- Conditions required for safe reset of the SCAI

Typically, SCAI have sensors to detect the process condition, a logic solver to determine when to take action, and final elements to take action on the process. The decision and action can be manual (e.g., operator response to a safety alarm) or automatic (e.g., via an interlock or SIS). In either case, the operator should understand when the SCAI take action, how the action affects the process, what the operator is expected to do in response to the action, and what the operator should do when a fault is detected in the SCAI. This information should be included in operating procedures as appropriate.

Each SCAI device has distinct failure modes and a failure rate that contributes to the overall SCAI performance. To be counted as IPLs, SCAI are designed and managed to achieve at least an order of magnitude of risk reduction (PFD $\leq$ 0.1). SCAI implemented in BPCS equipment are generally limited to a claim of one order of magnitude of risk reduction (PFD = 0.1) due to the typical BPCS design architecture and management systems. The documentation and validation requirements in the data tables in this section provide screening criteria for evaluating SCAI for risk reduction. Higher claims (e.g., PFD < 0.1) are supported by a SIS designed and managed in compliance with IEC 61511 (2003).

It is a good practice to implement SCAI equipment such that it takes the safe state action on failure rather than failing to a dangerous state. Examples of failures to consider are loss of signal, out-of-range values, loss of communications, power failure, instrument air failure, and loss of other utilities. If it is intended to continue operation of the process with out-of-service SCAI equipment, the risk of this temporary operation is assessed and compensating measures are implemented to address the risk. Out-of-service periods are tracked and minimized to the degree possible through equipment design and maintenance practices.

Validation is performed prior to the startup of any process with a new or modified SCAI. Validation demonstrates that the installed SCAI operates according to the specification and that procedures and documentation are in place to support its long-term management.

An equipment list should be maintained that identifies SCAI equipment by a unique designation, such as a tag number, that is traceable to the ITPM requirements necessary to ensure the equipment integrity and reliability throughout its life. The ITPM program includes a variety of activities, such as inspection, calibration, preventive maintenance, repair/replacement, and proof testing.

Periodic ITPM is required for SCAI to ensure equipment integrity and reliability throughout its life. Inspections and proof tests are conducted at intervals necessary to achieve the claimed PFD. Visual inspection may be more frequent in severe service conditions. Inspections and proof tests are performed according to a procedure by personnel who are capable of identifying the as-found/as-left conditions and verifying the integrity and reliability of the equipment. Records include as-found/as-left conditions, the tester, when tested, the procedure and equipment used, and calibration records.

The actual level of risk reduction achieved is generally substantiated by ITPM and human reliability data. The records associated with SCAI are used to confirm that the equipment operates as specified during all intended operating modes. Failure tracking and analysis is important to verify LOPA assumptions and to support continuous improvement. Historical records or evidence of prior use serve as demonstration that equipment is providing the required risk reduction.

As with other IPLs, changes to SCAI are reviewed using MOC, and access to SCAI equipment is controlled. Any change potentially affecting SCAI is assessed to determine the requirements for proof testing the SCAI functionality following the change. Access to engineering interfaces is also controlled to reduce human error and prevent compromise of the system. These measures are needed to reduce the potential that human errors could disable or degrade SCAI.

When programmable electronic systems (PES) are used, MOC and access security becomes more challenging because there are typically multiple functions within the common system. The software employed also makes changes easier to carry out and harder to detect. Human error potential increases because the individual who is granted system privilege has access to multiple functions at the same entry point. For example, an MOC that approves the change to one line of programming grants access to the entire program. The potential for human error also grows with the system size and complexity. Changes to embedded software in SCAI require complete functional testing of the hardware and application software. Changes to the application software are examined to determine the impact of the change. Any impacted logic is function tested to ensure its correct operation.

SCAI are sufficiently robust to withstand environmental stresses and provide the required integrity and reliability. Some technologies used in control applications may not be acceptable for SCAI due to inadequate performance history, inadequate reliability/integrity, or unpredictable failure behavior. For example, at the time of this publication, wireless technology as defined by ISA 100.11a (2011) is not generally considered an acceptable technology for executing SCAI but may be used for monitoring, status, or diagnostic communication.

5.2.2.1.1 Types of SCAI Systems

SCAI are considered complex systems because multiple components are required for successful operation. The SCAI boundary includes the instrumentation and control equipment, wiring, human machine interface (HMI), internal and external data communication, power supply, utilities (such as instrument air or hydraulics), and humans that act in response to issued alerts and alarms (Figure 5.9).

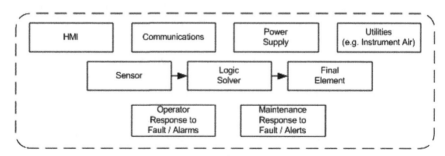

Figure 5.9. SCAI boundary.

LOPA is used to identify functions that are necessary to reduce the likelihood of a scenario. For SCAI, the identified function is allocated to a system that is designed and managed in a manner that supports the particular function and provides the required risk reduction. The instrumented system often supports (or executes) other tasks, such as nonsafety functions, diagnostics, reset, maintenance and testing, and manual shutdown. Fundamentally, the capability of any SCAI to stop a scenario is constrained by the design and management of the instrumented system that executes it.

In LOPA, it is assumed that each IPL is executed by a separate and independent system. The function may be performed by a single loop controller or by an integrated system. In practice, the instrumented function is referred to as a loop to emphasize that the equipment boundary includes all of the devices used to accomplish a specific safeguarding function. When multiple functions (or loops) are implemented in an integrated system or multiple systems are interconnected, any shared equipment should be evaluated for impact to the overall claimed performance. These common components can limit the achieved risk reduction to much less than the simple LOPA math would indicate.

In this section, SCAI are divided into four types of systems: alarms, controls, interlocks, and SIS. During LOPA, specific loops within these systems are identified, and the design verification establishes that these loops are capable of providing the required PFD. Any lack of independence between the systems should be assessed to ensure that the risk reduction credit taken for the instrumented systems is appropriate given any similar or shared hardware, procedure, or ITPM program elements.

Safety Alarm Loop

Description: A safety alarm loop uses instrumentation and controls to initiate an alarm, relies on human action to respond to the alarm, and includes the interfaces and final control elements used by the operator to take the required action. Example configurations of safety alarm loops using BPCS or SIS equipment are shown in Figure 5.10. Refer to ANSI/ISA 18.2 (2009) and IEC 61511 (2003) for additional guidance related to the instrumentation and control design.

Changes to the safety alarm loop hardware and software are controlled using an MOC process. During any process operating mode where the scenario could occur, the safety alarm loop is only bypassed (suppressed or setpoint changed) after administrative approval and the implementation of any necessary compensating measures. Administrative approval may involve formal MOC, implementation of bypass procedures with event tracking, following operating procedures, etc.

A safety alarm loop alone is not an IPL since the alarm does not directly take action to return the process to a safe state. A human response is required as part of the IPL. There are additional considerations associated with the human response portion of the IPL that are discussed in Section 5.2.2.5. A generic PFD value of a human response IPL, using either a safety alarm or another indicator as a trigger to respond, is given in Data Table 5.46.

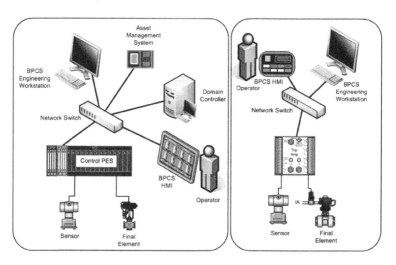

Figure 5.10. Examples of safety alarm loops using a BPCS and a SIS (Summers 2014).

Safety Control Loop (Normally Operating Control)

Description: A safety control loop includes instrumentation and controls that normally operate to support process (or regulatory) control and, by means of this operation, act to stop scenario progression related to other failures. Since these controls are not typically designed or managed in accordance with IEC 61511 (2003), they usually do not meet the requirements of SIS IPLs. The risk reduction claimed for a BPCS loop (which does not conform to IEC 61511 [2003]) must not be greater than 10 (IEC 61511 Clause 9.4.2). In LOPA, a risk reduction factor of 10 or PFD = 0.1 is used as an order of magnitude value, when the requirements are met for the BPCS loop to be claimed as an IPL (i.e., independence, reliability, etc.).

Safety control loop operation may be continuous (e.g., PID control) or intermittent (e.g., batch sequencing) in responding to process conditions within the normal operating range. For the safety control loop, the process demand is typically a change in process conditions within the normal operating range that results in the safety control loop taking action to keep the process within the operating envelope.

As an example, a temperature control loop may normally operate to control temperature in a vessel. Excessive heat could result in a pressure increase that could exceed the equipment rating. A loss of cooling water supply to the unit could prevent effective temperature control. A BPCS pressure control loop is in place to vent excess pressure and prevent vessel overpressure. If this operation is sufficient to prevent the vessel pressure from exceeding the safe operating limit, the pressure control loop may be a valid IPL.

Consequences Avoided: The safety control loop prevents progression of a scenario to a consequence of concern following an initiating event.

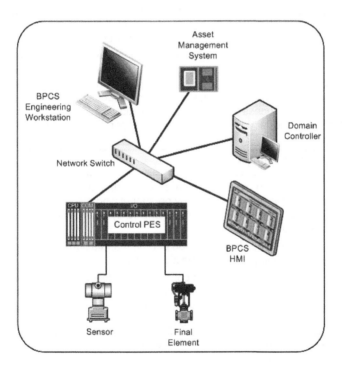

Figure 5.11. Example of safety control loop using a BPCS (Summers 2014).

Data Table 5.12. Safety control loop

IPL description
Safety control loop
Generic PFD suggested for use in LOPA
0.1
Special considerations for use of generic PFD for this IPL
A safety control loop consists of a sensor, controller, final control element, and support utilities and interfaces. It normally operates to support process (or regulatory) control. Its operation may be continuous or intermittent in responding to process deviations within the normal operating envelope, and its actions are sufficient to achieve or maintain a safe state of the process. • Equipment (or loop) failures may be revealed through process operation, automated diagnostics, or ITPM activities. The risk of continued process operation with a detected failure of the safety control equipment is assessed, and compensating measures are implemented to address any increased risk. • Changes to the safety control loop hardware and software are controlled using an MOC process. During any process operating mode where the scenario could occur, the safety control loop is only bypassed (placed into manual control) by procedure. Since manual operation depends on operator response to changes in the process variable, refer to Section 5.2.2.5 for guidance on considering operator actions as IPLs if manual operation of the safety control loop is permitted.
Generic validation method
• Inspection, testing, and maintenance of safety control loop equipment is performed according to a procedure by personnel who are capable of identifying the as-found/as-left conditions and verifying the integrity and reliability of the equipment. • The inspection frequency is determined per the manufacturer's recommendations and previous inspection history. • Documents include as-found/as-left conditions, the tester, when tested, the procedure and equipment used, and calibration records. • Periodic audits of auto/manual selection or bypassing records can indicate if the intended function has previously been impaired.
Basis for PFD and generic validation method
Consensus of the *Guidelines* subcommittee. See ANSI/ISA 84.00.01-2004 (IEC 61511 Mod) (ANSI/ISA 2004.) and *Guidelines for Safe and Reliable Instrumented Protective Systems* (CCPS 2007b).

Safety Interlock

Description: A safety interlock consists of instrumentation and controls that take automatic action to achieve or maintain a safe state of the process in response to a deviation from normal operation. The safety interlock may be implemented in a single loop controller, discrete control system (e.g., on/off, relay), distributed control system, programmable logic controller, or safety controller.

The risk reduction claimed for a safety interlock implemented in a BPCS (that does not conform to IEC 61511 [2003]) must not be greater than 10 (IEC 61511 Clause 9.4.2). In LOPA, a risk reduction factor of 10 or PFD = 0.1 is used as an order of magnitude value. When it is designed and managed per IEC 61511 (2003), the safety interlock is a SIS loop.

As an example, a temperature control loop may normally operate to control temperature in a vessel. Excessive heat could result in a pressure increase that could exceed the equipment rating. A loss of cooling water supply to the unit could prevent effective temperature control. A safety interlock is in place to automatically vent the vessel through a block valve and prevent overpressure. If this action is sufficient to prevent the vessel pressure from exceeding the safe operating limit, the safety interlock may be a valid IPL.

Consequences Avoided: Safety interlocks prevent progression of a scenario to the consequence of concern following an initiating event.

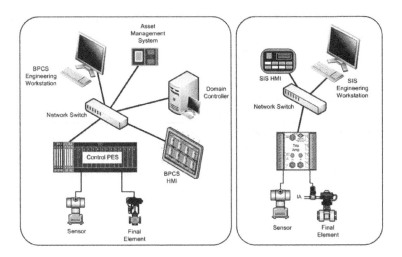

Figure 5.12. Examples of safety interlocks using BPCS and SIS (Summers 2014).

Data Table 5.13. Safety interlock

IPL description
Safety interlock
Generic PFD suggested for use in LOPA
0.1
Special considerations for use of generic PFD for this IPL
A safety interlock consists of a sensor, controller, final control element, and support utilities and interfaces. It acts to achieve or maintain a safe state of the process in response to a deviation from normal operation. • Equipment (or loop) failures may be revealed through process operation, automated diagnostics, or ITPM activities. The risk of continued process operation with a detected failure of the safety interlock equipment is assessed, and compensating measures are implemented to address any increased risk. • Changes to the safety interlock hardware and software are controlled using an MOC process. During any process operating mode where the scenario of concern could occur, the safety interlock is only bypassed with administrative approval and the implementation of any necessary compensating measures. Administrative approval may involve formal MOC, implementation of bypass procedures with event tracking, following operating procedures, etc.
Generic validation method
• Inspections and proof tests are conducted at an interval necessary to achieve the claimed PFD. • Inspection and proof testing are performed according to a procedure by personnel who are capable of identifying the as-found/as-left conditions and verifying the integrity and reliability of the equipment. • Documents include as-found/as-left conditions, the tester, when tested, the procedure and equipment used, and calibration records.
Basis for PFD and generic validation method
Consensus of the *Guidelines* subcommittee. See ANSI/ISA 84.00.01-2004 (IEC 61511 Mod.) (ANSI/ISA 2004) and *Guidelines for Safe and Reliable Instrumented Protective Systems* (CCPS 2007b).

Safety Instrumented System (SIS) Loop

Description: The SIS loop consists of instrumentation and controls designed in accordance with IEC 61511 (2003) that take action to achieve or maintain a safe state of the process after a process demand has occurred. Unless it is intended to implement the equipment within the BPCS according to IEC 61511 (2003), the equipment executing the SIS loop must be independent and separate from the BPCS equipment to the extent that the safety integrity of the SIS is not compromised (IEC 61511 Clause 11.2.4). SIS loops are typically designed to operate in low demand mode, but may operate in high demand (or continuous) mode due to frequent challenges from the process or its equipment. Refer to Sections 3.5.1 and 3.5.2 for more guidance on the consideration of mode of operation in LOPA.

IEC 61511 (2003) provides an extensive set of requirements covering the SIS lifecycle, from process hazard analysis through decommissioning. Some requirements can be very detailed and difficult to meet without special design practices, so the standard should be reviewed and understood in detail. For example, when PES are used in the SIS, diagnostic algorithms are required to detect input, output, and main processor failure; and to ensure that they are configured to fail to the safe state.

While it has many prescriptive requirements, IEC 61511 (2003) is fundamentally a performance-based standard that requires the establishment of a performance benchmark, the safety integrity level (SIL), that the SIS loop must meet. The SIL needed by the SIS loop is normally established using a risk analysis technique such as LOPA. For example, a LOPA can determine the PFD that the SIS loop is required to achieve to meet a facility's risk tolerance criteria. The required PFD establishes the target SIL for the SIS loop. Table 5.1 shows the relationship between SIL and the PFD ranges for low demand mode SIS loops.

Table 5.1. Safety integrity level relationship to PFD and risk reduction

SIL Level	PFD Is at Least:	But Less than:	Typical Risk Reduction Credited
SIL 1	0.01	0.1	1 order of magnitude
SIL 2	0.001	0.01	2 orders of magnitude
SIL 3	0.0001	0.001	3 orders of magnitude

When claiming SIL 2 and 3, the analyst should consider that these are claims of PFD = 0.01 to 0.001 and PFD = 0.001 to 0.0001, respectively, for a single IPL. Depending on the required SIL, the standard also has minimum redundancy, or fault tolerance, requirements to ensure adequate protection against systematic errors. Redundant devices and signal paths using independent sensors, controllers, or final elements to provide the same function may be necessary to achieve SIL 2, and are required to ensure no single points of failure for SIL 3.

The SIS equipment is covered by rigorous access controls, and there is a program for ongoing proof testing. Systematic errors and failures should be considered during verifications and functional safety assessments; these errors might occur during design, operation, maintenance, or validation. These management systems, as well as other measures and constraints, are intended to reduce the systemic errors that would otherwise limit the performance of the system as designed, installed, operated, tested, and maintained.

The PFD of the SIS loop is verified by quantitative analysis of the architecture using failure rate data based on the capability of the SIS equipment as installed in the operating environment (IEC 2003). The PFD should be demonstrated to be at least as good as what is claimed in the LOPA. In addition, the SIS should be evaluated for compliance with the detailed requirements of IEC 61511 (2003), especially those with regard to separation from the BPCS, fault tolerance, verification, and validation.

The ISA84 committee has developed a series of complementary technical reports to provide guidance and practical examples related to various SIS topics and applications. Three of these technical reports, ISA-TR84.00.02 (2002), ISA-TR84.00.03 (2012), and ISA-TR84.00.04 (2011), provide a comprehensive overview of the SIS lifecycle. Data Table 5.14 provides appropriate screening criteria for identifying SIS, although it is not intended to be a substitute for meeting the requirements of IEC 61511 (2003).

Consequences Avoided: A SIS loop prevents progression of a scenario following an initiating event.

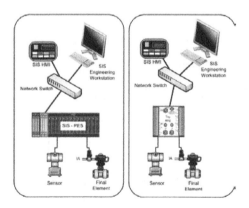

Figure 5.13. Examples of SIS using a PES and a trip amplifier (Summers 2014).

Data Table 5.14. SIS loop

IPL description
SIS loop
Generic PFD suggested for use in LOPA
SIL 1 – 0.1 SIL 2 – 0.01 SIL 3 – 0.001
Special considerations for use of generic PFD for this IPL
SIL refers to the performance achieved by the SIS loop given its probability of random and systematic failure in the operating environment. Each SIS loop achieves or maintains a safe state of the process with respect to a specific scenario. Sensors, logic solvers, final elements, and interconnected equipment that are required for the SIS loop to operate according to its specification are considered in the PFD calculation. The achievement of a PFD of < 0.1 requires rigorous design and management practices to ensure that the SIS loop is capable of achieving the claimed risk reduction in the specific operating environment, including sufficient protection against systematic and common cause effects. Unless it is intended to implement the equipment within the BPCS per IEC 61511 (2003), the equipment executing the SIS loop must be independent and separate from the BPCS equipment to the extent that the safety integrity of the SIS is not compromised (IEC 61511 Clause 11.2.4). *For the PFD claimed above to be valid for SIL 2 and SIL 3 systems, the design of human interfaces for operations and maintenance should minimize the potential for human error during installation, maintenance, testing, and bypassing.* • Equipment (or loop) failures may be revealed through process operation, automated diagnostics and ITPM activities. The risk of continued process operation with a detected failure of the SIS equipment is assessed, and compensating measures are implemented to address any increased risk. • Changes to the SIS loop hardware and software are controlled using an MOC process. During any process operating mode where the scenario could occur, the SIS loop is only bypassed with administrative approval and the implementation of any necessary compensating measure. Administrative approval may involve formal MOC, implementation of bypass procedures with event tracking, following operating procedures, etc. • The design, operation, configuration management, and ITPM practices ensure that the actual performance of the installed SIS loop achieves the target SIL.

Table 5.14. Continued

Generic validation method
• Inspection and proof testing are performed at an interval necessary to achieve the claimed PFD and according to a procedure.
• The procedure should be performed by personnel who are capable of identifying the as-found/as-left conditions and verifying the integrity and reliability of the equipment.
• Documentation includes as-found/as-left conditions, the tester, when tested, the procedure and equipment used, and calibration records.
Basis for PFD and generic validation method
Consensus of the *Guidelines* subcommittee. See ANSI/ISA 84.00.01-2004 (IEC 61511 Mod.) (ANSI/ISA 2004) and *Guidelines for Safe and Reliable Instrumented Protective Systems* (CCPS 2007b).

5.2.2.1.2 Counting Multiple BPCS Loops in One BPCS in the Same Scenario

In *CCPS LOPA* (2001), two approaches were presented for considering the BPCS in LOPAs. The first approach, called "Approach A," assumes that a single BPCS CPU failure invalidates the use of all other BPCS loops using the same CPU. The key criteria for a BPCS IPL in Approach A are: (1) the IE is not related to equipment within the BPCS, and (2) the BPCS IPL is properly designed and managed in order to claim one order of magnitude of risk reduction. The second approach, "Approach B," potentially allowed two BPCS loops on the same CPU to be credited for the same hazard scenario. Since the 2001 *CCPS LOPA* edition, additional thoughts and standards have been developed with respect to IPLs implemented using BPCS equipment.

This section provides guidance and criteria for evaluating the appropriateness of Approach B for a particular application and describes the rigor of the analysis and proof that is required to support the IPL claims. The typical design and management practices associated with BPCS equipment limit its capability to one order of magnitude per logic solver (e.g., controller) (IEC 2003). When claiming more than one order of magnitude from a single controller in a scenario, IEC 61511 (2003) requires that the controller be designed and managed as a SIS in accordance with IEC 61511 (2003).

There are many important requirements that are necessary to ensure that the less conservative Approach B achieves the claimed risk reduction. The reader is urged to read, understand, and meet these requirements when crediting two BPCS loops on a common CPU in the same hazard scenario.

It is recommended that users who choose Approach B consider the use of advanced techniques, such as fault tree or other quantitative risk assessment techniques, to ensure that potential common cause failures (including systematic errors) are fully assessed. Since a common cause or systematic failure could result in the simultaneous loss of control and safety protection, the LOPA team should carefully examine the independence and separation of the two systems.

CAUTION:

An updated version of the IEC 61511 standard is currently being developed. The standards committee is considering limiting both the total number of BPCS credits and the total number of instrumented credits that may be claimed in a single hazard scenario. The reader is advised to maintain an awareness of changing standards that may have an effect on an organization's LOPA practices.

Special consideration should be given to situations where low demand mode functions are implemented in BPCS equipment. It is inherently more reliable to implement low demand mode functions as safety interlocks in SIS equipment, since the SIS equipment is specifically designed and managed for that purpose.

Further, certain application-specific practices require that safety interlocks be implemented in an independent SIS. Where other recognized practices apply, they are followed regardless of the rules adopted in a particular LOPA procedure.

Comparison of Current Approaches

Approach A allows the identification of a BPCS as either an IE or an IPL (but not both), while Approach B provides the basis for a second order of magnitude of risk reduction (an IE and IPL or 2 IPLs) on a single CPU using additional hardware and more rigorous practices. Approach A takes the conservative position that any failure of BPCS equipment invalidates all other loops implemented within the BPCS. It is used for LOPA because its rules are clear and conservative. It provides a high level of protection against common cause failures between the IE and the IPL or between 2 IPLs. Approach B assumes that the two BPCS loops are designed and managed such that the likelihood (probability) of simultaneous failure is low enough to support an estimated failure rate of ≤ 0.01/yr (IE and IPL) or two orders of magnitude of risk reduction (2 IPLs). The use of Approach B requires that the analyst is experienced in BPCS design, has adequate data available on the actual performance of the BPCS, and understands how to identify and account for common cause failures. Approach B also requires management commitment to enforce the rigorous practices necessary to control systemic, common cause, and dependent errors to a sufficiently low level.

Performance Considerations in Using Approach B

Each device in a BPCS loop has its own failure rate as a result of its design, manufacture, installation, ITPM program, etc. The loop performance is related to the hardware, software, and systematic failures that can cause the BPCS loop to not operate as required. The estimation of its performance requires consideration of the device failure rates, the system architecture, and any systematic errors, such as software or human errors. Verification can be done with any of the approaches discussed in Chapter 3; for instance, for the site-specific data approach, the ITPM records can be used to demonstrate that the devices are achieving the required performance.

The risk reduction capability of the BPCS loop is limited by its hardware and software design. Most commercial BPCS logic solvers and alarm systems have little online redundancy in the input/output cards (I/O) or in the CPU. When redundancy is available, it is generally a backup processor that does not operate until internal diagnostics (typically not themselves redundant) detect a problem (e.g., hot standby or hot swappable processor). This limits the CPU performance.

The internal diagnostics are also limited in coverage in most commercial BPCS logic solvers. In contrast, a PES used in SIS is required to have diagnostics on input and output cards to detect stuck-on or stuck-off contacts and on the main processor to detect a stuck or locked-up processor. These diagnostics are necessary for the PES to have a performance level that meets the reliability requirements for a SIS loop with a specified SIL. These diagnostics are not required for hardwired logic solvers, such as relays or trip amplifiers, which have low failure rates and well-defined failure modes. Hardwired logic solvers are separate and diverse from other control systems and are not vulnerable to cyber-security risks or data highway corruption.

Correct operation of instrumentation and control systems requires multiple devices to function correctly. The cumulative effect of each device's potential for failure often results in the need for individual devices to achieve an installed performance at least an order of magnitude better than the required system performance. For example, if it is desired to claim that a BPCS loop achieves a $PFD = 0.1$, the BPCS controller would typically need to have a $PFD \leq 0.01$. Industry and manufacturer data indicate a random hardware failure rate for typical BPCS hardware (such as DCSs/PLCs) in the range of 10^{-5}/hr to 10^{-6}/hr (ISA TR84.00.04-2011 Appendix F.2.1 [ISA 2011]). This failure rate range indicates that typical BPCS are limited to a $PFD = 0.1$ claim.

When two loops share any device, including the logic solver or CPU, the system created by the two loops must be assessed and determined to have a failure rate ≤ 0.01/yr to justify the performance claim for the two loops. This determination includes consideration of possible common cause failures between the devices of the system and the greater potential for systematic errors. Site-specific analysis is necessary to substantiate a failure rate claim of < 0.1/yr for the overall system, if it is not designed and managed in compliance with IEC 61511 (2003).

Similar to IEC 61511 (2003), an Approach B design provides independence and fault-tolerance, backed with a robust management system that includes documentation, access controls, verification, validation, and auditing. Rigorous design and management may decrease the potential for BPCS failure; however, it also increases administrative costs and constrains the ability of the operators and process engineers to interact with, and to optimize, the control parameters and applications that may be on the same logic solver as the BPCS loop.

In many cases, a BPCS logic solver requires input from humans to perform its normal operating tasks, increasing the potential for human error. Often, logic solvers within BPCSs are deliberately made accessible to personnel who have the ability to change set-points, bypass alarms, place control loops in manual mode, make program changes, etc. Most BPCS logic solvers do not have built-in version tracking capability, making it difficult to detect configuration (or program) changes. In many cases, embedded software (e.g., firmware) and application program changes are made without verification and validation.

BPCS equipment is being increasingly integrated with plant and other external systems, such as business enterprise software systems, manufacturer maintenance management systems, remote links for central engineering access, or at-home access, etc. This openness has brought great benefits to the process industry, but it does make protection layers within the BPCS more vulnerable to human error and deliberate acts of sabotage.

Therefore, using Approach B requires rigorous management of change and access control to ensure cyber security and software/data integrity. In some installations, it might be possible to impose sufficient control over access and MOC requirements on BPCS logic solvers. In others, this could impose unacceptable operational constraints. Before using Approach B, it is important to ensure that there is sufficient analysis and test data to demonstrate that the BPCS for a particular process is designed and managed in a way that the two BPCS loops, in combination, can achieve the overall performance claim of < 0.01/yr.

At a minimum, the system analysis includes the following:

- Assessment of potential common cause, common mode, and systematic failures between the BPCS IE and IPL, or the two BPCS IPLs, to determine that their impact is sufficiently low as compared to the performance claim.

- Written specification covering the safety loops within the system, such as instrument diagrams, P&IDs, loop diagrams, and functional specifications.

- For the generic data approach to validation, the assessment of previous historical performance of the BPCS CPU, input/output cards, sensors, final elements, human response, etc. *(NOTE: manufacturers' information should be examined critically to ensure that it applies to situations similar to a particular installation.)*

- For the site-specific data approach to validation, evaluation of inspection, maintenance, and test data over a significant period to demonstrate that the system achieves the performance claim.

- Assessment of access security measures for hardware and software.

- Management of change and revision control for the hardware and software, including setpoints, configuration, and operator overrides.

This analysis requires more expertise and a more detailed understanding of the BPCS hardware and software design than is typically present on LOPA teams. Consequently, additional verification and functional assessment by a person competent in such analysis are required to ensure that the BPCS integrity and reliability are sufficient.

Design Considerations in Using Approach B

CCPS LOPA (2001) stated that no more than a total of two BPCS loops sharing a common CPU should normally be credited for the same scenario due to potential common cause failure and systematic error. Restrictions are also included in IEC 61511 (2003). Clause 8.2.2 addresses the BPCS as an IE and Clauses 9.4.2 and

9.4.3 discuss the use of the BPCS as part of an IPL. These restrictions take into account two aspects of the BPCS:

- The random failure rate of hardware that is not designed and managed in compliance with IEC 61511 (2003)
- The systematic error associated with typical BPCS design and management

When claiming credit for a second BPCS loop on a single CPU in the same LOPA scenario (either the BPCS as the IE and as an IPL, or as two BPCS IPLs if the IE is not the BPCS), the following should be confirmed:

- If the two BPCS loops share one CPU, the design and performance considerations for Approach B, previously discussed in Section 5.2.2.1.2, are met.
- The sensor is independent of the sensor that is part of the IE and any other IPL credited for the scenario.
- The final element is independent of the final element that is part of the IE and any other IPL credited for the scenario.
- Access controls and management of change procedures reduce the potential for human error (e.g., incorrect configuration, program error, unapproved bypassing) to a sufficiently low level as compared to the claimed performance.

In addition, when using the site-specific data approach to validation, sufficient actual test data, calibration data, and failure data should be collected and the data should be of sufficient quality to validate that the PFD is achieved.

Figure 5.14 illustrates the use of two independent controllers as shown in IEC 61511 (2003) Part 2, Figure 2. This example configuration does not incorporate any communication (or interconnection) between the two BPCS loops. This configuration is an inherently more reliable design, as it is physically separate as well as independent, so it provides a high degree of protection against data corruption and inadvertent software changes. The segregation of the control and safety I/O also provides higher reliability of the SCAI.

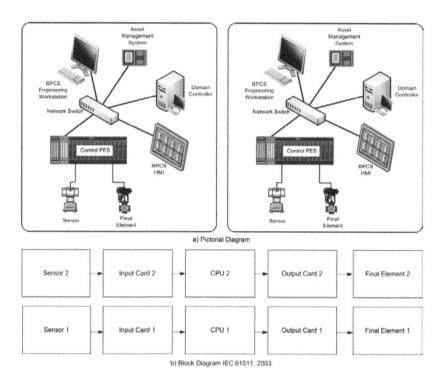

a) Pictorial Diagram

| Sensor 2 | → | Input Card 2 | → | CPU 2 | → | Output Card 2 | → | Final Element 2 |

| Sensor 1 | → | Input Card 1 | → | CPU 1 | → | Output Card 1 | → | Final Element 1 |

b) Block Diagram IEC 61511, 2003

Figure 5.14. Two BPCS loops with independent controllers credited for the same scenario (Summers 2014).

Guidelines for Safe and Reliable Instrumented Protective Systems (CCPS 2007b) recognized that the controllers could be functionally separate while safely sharing information. Figure 5.15 illustrates this configuration, which consists of separate controllers that communicate in a secure manner using a common data highway. This system has the potential to achieve a 0.01 PFD claim for the two BPCS loops if the system is properly designed and managed, and each BPCS loop meets the conditions described for safety control, safety alarm, or interlock as appropriate. A BPCS architecture that uses a primary and hot backup controller is not sufficient to be credited as two independent controllers due to potential common cause failures. For example, a hot backup controller shares common equipment with the primary controller, such as the backplane, firmware, diagnostics, and transfer mechanisms.

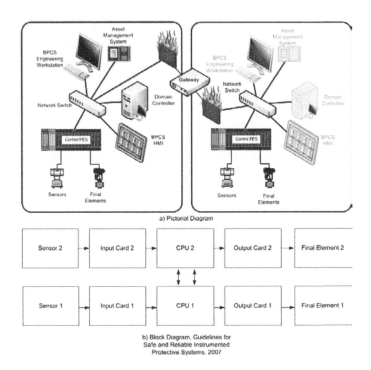

a) Pictorial Diagram

| Sensor 2 | → | Input Card 2 | → | CPU 2 | → | Output Card 2 | → | Final Element 2 |

| Sensor 1 | → | Input Card 1 | → | CPU 1 | → | Output Card 1 | → | Final Element 1 |

b) Block Diagram. Guidelines for
Safe and Reliable Instrumented
Protective Systems, 2007

Figure 5.15. Two BPCS loops with independent controllers and secure communication credited for the same scenario (Summers 2014).

Another BPCS configuration is shown in Figure 5.16. This configuration is adapted from Figure 11.5 in *CCPS LOPA* (2001). This architecture does not generally support the crediting of two independent BPCS loops.

CCPS LOPA (2001) indicated that the BPCS CPU had at least two orders of magnitude (or better) performance than the field devices. In the decade since that publication, data have not been produced to support this position. Data indicate that the failure rate of a typical BPCS CPU is >0.01/yr (*PDS Data Handbook* [SINTEF 2010]), so the typical BPCS CPU performance is equivalent to many other electrical, mechanical, or programmable electronic devices. Therefore, it will be difficult to validate that the configuration in Figure 5.16 qualifies for Approach B because Approach B claims that a total failure rate of the two loops taken together would be ≤ 0.01/yr.

While the field devices are connected to separate input and output cards, the logic processing is executed by a shared CPU. Since the BPCS CPU is a single point of failure, the CPU must demonstrate a failure rate that is lower than 0.01/yr in order for the overall BPCS created by the two loops to achieve a failure rate <0.01/yr. As discussed earlier, BPCS logic solvers generally have failure rates that are >0.01/yr, so they do not support two BPCS loops for the same scenario.

BCPS loops can be implemented using interconnected systems protected by firewalls that secure safety data and logic or as independent DCSs, PLCs, alarm systems, or single loop controllers. When two BPCSs share equipment or support systems that can cause the dangerous failure of the interconnected BPCSs, the resulting interconnected system should be analyzed as a single system. If the risk reduction claimed for a BPCS is >10, IEC 61511 (2003) Clause 9.4.3 requires that the BPCS be designed and managed as a SIS.

> Note: Manufacturer claims regarding individual devices are not sufficient to justify lack of independence between the two BPCS IPLs or between a BPCS IPL and a BPCS IE for the same scenario.

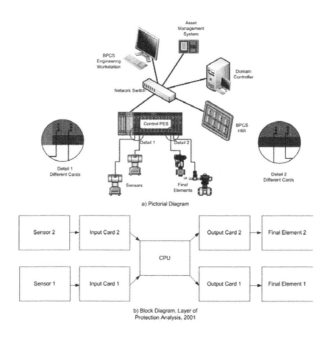

Figure 5.16. Two BPCS loops with a shared CPU typically cannot support multiple BPCS loop credit for the same scenario (Summers 2014).

The configuration as shown in Figure 5.17 achieves even worse performance (higher failure rate or less risk reduction) than Figure 5.16 due to the potential common cause failures associated with the BPCS CPU and I/O cards. This architecture is generally not acceptable for crediting two BPCS loops. This is consistent with *CCPS LOPA* (2001), which recommended against taking credit for a second function in the BPCS where an input or output card is common between the loops.

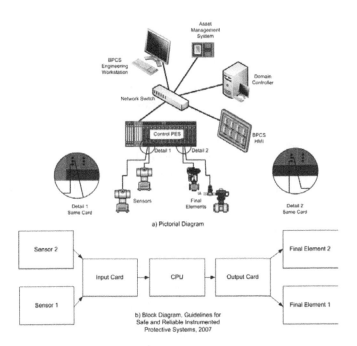

Figure 5.17. Two BPCS loops with a shared CPU and shared I/O cards does *not* support multiple BPCS loop credit for the same scenario (Summers 2014).

Essentially, the overall performance of the two loops illustrated in Figures 5.16 and 5.17 is limited by the integrity of the BPCS CPU, which is generally designed to achieve a failure rate between 0.1/yr and 0.01/yr. Taking credit for two loops in either configuration requires that the system be designed and managed according to IEC 61511 (2003). Quantitative verification of the system (or of a similar baseline system that is used as a model for other systems) is necessary to ensure the hardware integrity and a qualitative evaluation is used to identify sources of systemic error so that common cause is properly addressed in the design, validation, and management of the system.

5.2.2.1.3 Crediting Shadowed Interlocks in the BPCS

Shadowing was a common practice within the chemical industry prior to the widespread availability of safety-configured PLCs designed in compliance with IEC 61508 (2010). Shadowing involved the replication of the SIS logic in the BPCS logic. Shadowed logic in the BPCS logic solver uses a BPCS sensor but may also use the SIS sensor as well. Upon sensing the specified process condition, the shadowed logic actuates the BPCS final control element and may also actuate the SIS final element.

Initially, shadowing was used to address the potential for unrevealed failures when a general-purpose PLC was used as the SIS logic solver. General-purpose PLCs are industrial-grade logic controllers that are not safety-configured (e.g., they do not have watchdog timers, stuck-on/stuck-off diagnostics, redundant on-line channels, etc.). The shadowed logic within the BPCS equipment provided some protection against faults of the PLC hardware and errors in the embedded and application software.

Many companies choose to shadow the logic within the BPCS due to other benefits. Shadowing provides diversity of the embedded software and application program, reducing the potential that a systematic error in the SIS would allow a scenario to propagate. It also ensures that the BPCS logic solver takes its control elements to the shutdown condition to facilitate safe restart.

From a LOPA perspective, the risk reduction benefit of the shadowed function is limited, since the sensors and final elements are generally shared between the functions. If the analyst wishes to take risk reduction credit for the BPCS shadowed logic, quantitative risk assessment techniques can be used to properly account for the common cause failure potential. When credit is given to the shadowed logic, the risk reduction factor is generally significantly <10 due to common cause failure potential. Typically, this function is not listed as an IPL in LOPA due to the potential for common cause failures and errors between the BPCS and the SIS. Shadowed logic can provide a compensating measure during out of service periods for other IPLs.

5.2.2.1.4 Implementing Controls and Alarms in the SIS

Controls and alarms may be placed in a SIS, as long as they are managed according to the requirements of IEC 61511 (2003). Any equipment that is shared between the SIS loop and the control (or alarm) loop is designed and managed to meet IEC 61511 (2003) requirements. IEC 61511 (2003) provides metrics to support the failure analysis of the continuous and high demand mode system.

Significant issues can arise when controls are implemented in the SIS:

- Increased complexity results in demands for special resources to design and manage.
- Increased potential for common cause/common mode failures (hardware, software, and human).
- Limitation of SIS logic solver in performing the functions of both control loops and SIS loops (e.g., logic solver integrity, diagnostic requirements, processing speed, memory, self-checks, etc.).
- Limitation on control system software and configuration updates due to the need to protect the SIS from unapproved and unintended writes to memory and from unintended downloads (including protection from viruses).

- Added restrictions for changing control code since the MOC requirements for a SIS apply to the entire system.

Maintaining this required high degree of management system rigor can be quite challenging. As more non-safety functions are implemented in the SIS, there are more I/O to manage and a greater potential that access/changes to non-safety equipment could impact the SIS operation.

To minimize the potential for common cause failures and human error, some organizations mandate that controls reside in a BPCS logic solver and that IPLs be implemented in a separate and independent SIS; these users do not take credit for BPCS loops (controls, alarms, or interlocks) as IPLs. Some users choose to implement safety alarms in SIS equipment rather than in BPCS equipment, especially where they are the last line of instrumented defense that stops an incident scenario of severe consequence. Others implement safe operating limit alarms in a separate alarm system from the logic solvers used to implement the normal operating control loops and other SCAI.

5.2.2.1.5 Counting Multiple SCAI for the Same Scenario

Independence assessment is the examination of the equipment, procedures, and personnel that ensure SCAI performance. Whenever multiple SCAI are listed for the same scenario, there is the potential that common cause or systematic error may lead to the failure of multiple systems. This potential is also present when the IE involves instrumentation and controls and SCAI are used to stop event propagation. Common (or similar) elements between the IE and the listed SCAI indicate insufficient independence, which may be difficult to assess using the simple LOPA rules.

Three key questions should be considered when assessing the independence and risk reduction capability of SCAI:

1) Is there any piece of equipment being shared by the initiating event and any listed SCAI?

   Consideration should be given to the elements illustrated in Figure 5.9 and how their failure impacts the initiating event and listed SCAI.

2) Are there any equipment components used by the initiating event and by listed SCAI that are similar in design, installation, configuration or manufacture?

   Similar equipment introduces the potential for common cause failure that could result in the failure of multiple systems to operate when a process demand occurs. Consideration should be given to how common cause failure impacts the estimated hazard scenario frequency. For example, a scenario may be addressed by a safety alarm, which requires an operator to take action on the process based on high level, and a SIS, which takes the process to a safe state based on high-high level. If both SCAI use the same technology to detect level, there is the potential that both SCAI can be disabled by the same maintenance error or manufacturing problem.

3) Do the design and management practices ensure that personnel cannot while performing any activity inadvertently change, bypass, or disable any instrumentation that leads to the initiating event and the failure of the listed SCAI?

Personnel interact with, and potentially induce, errors that can impact any SCAI protecting a process. Common personnel or interfaces should be analyzed to determine whether the management system is strong enough to reduce the potential for systematic error to a sufficiently low level compared to the claimed risk reduction or hazard scenario frequency.

5.2.2.2 Relief Systems

Relief systems work by quickly opening sufficient venting area to relieve pressure or vacuum from a protected system, such that the system pressure is maintained in an acceptable range. The relief system typically includes one or more relief devices and can also include associated abatement systems (such as flares and scrubbers).

Relief devices have been used successfully by the chemical, petrochemical, refining, and power generation industries for over a century. Relief devices include relief valves, rupture disks, and other devices that provide a means for relieving pressure if it increases above a defined level. Conservation vents and vacuum breakers are often used to prevent the development of vacuum in a vessel. Different relief devices rely upon different physical principles for their successful operation. It is important to understand these principles and the potential failure modes of the devices under consideration to select a type with sufficient reliability for the specific application.

5.2.2.2.1 Impact of Isolation Valves on Relief Systems

Many individual relief devices that are part of the relief system IPLs can have a PFD of 0.01 or better. However, the *Guidelines* subcommittee was concerned that, unless the plant has a robust system for ensuring that block valves remain open, it would be inappropriate to assume that the error rate associated with an inappropriately closed in-line block is negligible. The conservative PFD of 0.1 is therefore used as the default value for relief devices that have block valves upstream or downstream of relief devices (unless the valves are 3-way valves that cannot impede flow in any valve position). Some organizations have robust management systems in place that ensure that block valves upstream or downstream of a relief device remain in their correct positions, and their management systems are periodically audited to ensure that the valve positions are appropriately controlled. Some companies also require x-ray analysis of valves that are located upstream or downstream of relief devices to determine if the relief paths are actually open and unimpeded. Where it can be shown that robust management systems are in place to ensure that control of valves are well managed, individual firms can claim a failure rate appropriate for their relief system performance.

5.2.2.2.2 Independence and the Relief Valve Design Basis

A basic tenet of LOPA is that IPLs are independent; that is, each IPL is capable of preventing the consequence of concern without relying on the operation of another IPL. It is important that the relief device be designed to handle the scenario of concern. There are, however, times when the relief device is sized using an assumption that another IPL will be at least partially effective. In this case, the relief valve effectiveness is dependent upon the success of another device, and the two components cannot be treated as separate IPLs.

Example: Low Pressure to High Pressure (LP/HP) Interface Protection

An example of the relief device design being dependent upon the successful functioning of another device can be found in low pressure to high pressure interface protection. When a low pressure system is connected to a high pressure system, protection of the low pressure system from overpressure may be required. Specific instances where this may be an issue include a

- Compressor system that boosts pressure from a low pressure system to a high pressure system
- Pump that sends liquid from a low pressure system to a high pressure system
- Vessel liquid seal that separates a low pressure system from a high pressure system

Of concern is the potential for reverse flow from the high pressure system to the low pressure system. The low pressure system may be overpressured by the high pressure system, resulting in a release of process material. To illustrate this concept, consider the compressor P&ID shown in Figure 5.18.

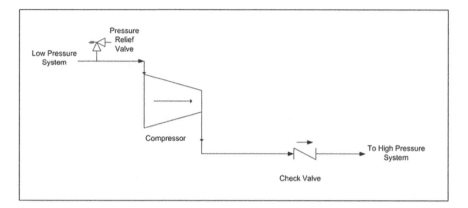

Figure 5.18. Low pressure to high pressure (LP/HP) system example.

It is a common practice to provide one or more check valves to prevent reverse flow from the high pressure system to the low pressure system. In addition, a pressure relief valve (PRV) is often provided to protect the low pressure system against a high pressure event. There are two different strategies that may be employed in selecting the design basis for the PRV capacity.

Case 1 – Design per ANSI/API 521 (2008) allowance for reverse flow through the check valve.

The guidance in ANSI/API 521 *Pressure-relieving and Depressuring Systems* (2008) states:

> *"Where no specific experience or company guidelines exist, one may estimate the reverse flow through series check valves as the flow through a single orifice with a diameter equal to one-tenth of the largest check valve's nominal flow diameter. A lower value may be used if a condition-monitoring system for the check valves (e.g. pressure indicators with appropriate volumes between the check valves) is installed to monitor the condition to ensure that the leakage rate is below the capacity of the low-pressure side relief device."*

If the ANSI/API 521 (2008) guidance is followed, the PRV sizing basis would be reduced to handle just the flow from a slight reverse leak through the check valve. If the check valve does not function when demanded (does not close or allows greater than design basis reverse flow), the PRV would be undersized and the low pressure system would be overpressured. Figure 5.19 presents an event tree that describes the various scenario paths that could arise when applying the ANSI/API 521 guidance. In this case, the proper functioning of the relief device is *dependent* upon the success of the check valve in substantially reducing the reverse flow. In this case, the relief valve is not independent of the check valve and would not be considered to be a separate IPL. The PFD for the check valve-relief device combination would be represented by the PFD of the IPL with the greatest probability of failure. In this case, the PFD of the check valve (PFD = 0.1) would represent the PFD of the overall system.

Case 2 – Design the PRV to handle the full, unrestricted reverse flow, assuming that the check valve fails to close.

In this design case, it is assumed that the check valve can fail completely open, and adequate pressure relief is provided for the low pressure system to protect against the full reverse flow from the high pressure system. Figure 5.20 presents an event tree that describes the various scenario paths that could arise using the Case 2 design basis. In this scenario, the relief valve design is *independent* of the check valve. If the check valve fails fully open, the relief valve is still capable of preventing system overpressure. In this scenario, the relief device could still qualify as an IPL, independent of the check valve.

Another example of dependency in relief valve sizing occurs when there is an assumption that cladding is in place to limit heat input to a vessel in the event of a fire. If the relief valve can only prevent the consequence of concern if the cladding is in place, then cladding would not be considered to be an independent IPL. However, if the relief valve is sized for the fire case without cladding, and if the cladding can reduce the heat flux to prevent the consequence of concern without requiring the operation of the relief device, then cladding could qualify as an IPL.

Figure 5.19. Event tree for a relief device sized for limited reverse flow.

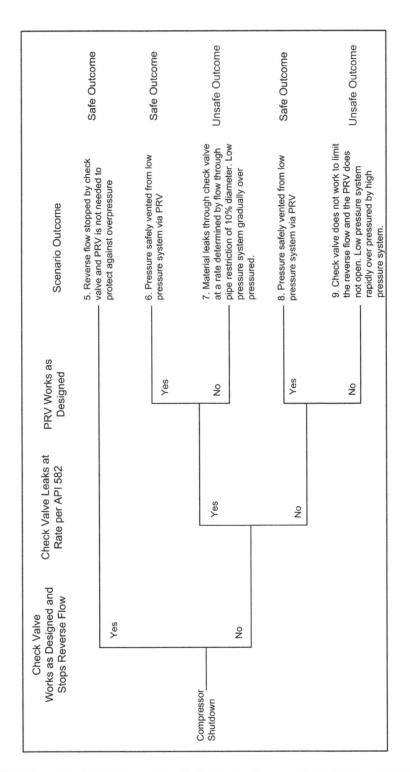

Figure 5.20. Event tree for a relief device sized for full reverse flow.

5.2.2.2.3 Relief Devices: Other Considerations

It is certainly important to ensure that the relief device is sized to prevent the scenario of concern, independent of the operation of other IPLs. For the relief valve to be an effective IPL, proper maintenance is essential. A number of countries and local jurisdictions have specific requirements concerning the frequency of inspections, the test methods to be employed, and the certifications of individuals who test relief devices. It is important to understand and follow the requirements of the individual jurisdiction.

A relief device can be an effective IPL to prevent overpressure; however, the subsequent release from the device can result in other consequences. Effluent release upon successful activation of a relief device can be evaluated as a separate LOPA scenario.

The following pages provide examples of various configurations of relief devices that are used in relief system IPLs. The IPL values suggested in the data tables are based on clean, nonfouling service and may not be appropriate for services that are prone to plugging or corrosion.

Data tables provided in this section include:

Data Table 5.15. Spring-operated pressure relief valve
Data Table 5.16. Dual spring-operated pressure relief valves
Data Table 5.17. Pilot-operated pressure relief valve
Data Table 5.18. Gas balance/adjustable set pressure surge relief valve
Data Table 5.19. Buckling pin relief valve
Data Table 5.20. Buckling pin isolation valve
Data Table 5.21. Rupture disk
Data Table 5.22. Spring-operated pressure relief valve with rupture disk
Data Table 5.23. Conservation vacuum and/or pressure relief vent
Data Table 5.24. Vacuum breaker
Data Table 5.25. Frangible roof

Spring-Operated Pressure Relief Valve

Description: These devices are held in the closed position by springs. When the pressure under the relief valve is great enough to overcome the force of the spring, the valve opens and passes material through the valve. Once the pressure under the valve has decreased, the spring forces the valve closed and the relief flow stops. Spring-operated pressure relief valves have proven to be highly dependable. However, valves that are installed in potentially fouling, polymeric, or freezing services may require more rigorous inspection and maintenance procedures and/or additional system design features, such as heat tracing, to maintain reliability.

Consequences: Typically, the consequence being analyzed in a LOPA scenario involving a relief valve is overpressure, resulting in a rupture of piping or a vessel. Generally, the onset of leakage of the piping or vessel is expected at pressures exceeding 121% of design. (Refer to Appendix E Considerations for Overpressure and Related Outcomes for more information.)

NOTE: If fire cladding was assumed to be in place on the protected vessel in the calculation of the size of a relief valve, then the combined PFD of the insulation plus relief valve is 0.01.

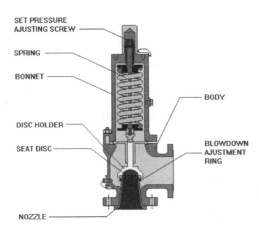

Figure 5.21. Spring-loaded pressure relief valve, courtesy of the University of Michigan

Data Table 5.15. Spring-operated pressure relief valve

IPL description
Spring-operated pressure relief valve
Generic PFD suggested for use in LOPA
0.01 for failure to open enough at set pressure (100% of rating) If there is an isolation valve (block valve) upstream or downstream of the relief device, then the suggested PFD is 0.1, unless there is a management system in place to ensure that valves are returned to service in their proper positions after maintenance and that they remain in the appropriate state during operation. *NOTE: If fire cladding was assumed to be in place on the protected vessel in the calculation of the size of a relief valve, then the combined PFD of the insulation plus relief valve is 0.01.*
Special considerations for use of generic PFD for this IPL
• The PRV is sized for the scenario being considered. • The inlet and outlet piping are sized correctly and are mechanically adequate for relief flow. • The relief valve is in clean service, and the metallurgy is corrosion-resistant to the particular service. • The service under evaluation does not have the potential for freezing of the process fluid before or during relief; if freezing is possible, then adequate heat tracing of the relief valve and piping is installed and maintained.
Generic validation method
• The ITPM frequency is set in accordance with the manufacturer's recommendations and/or code requirements and may be adjusted based on the results of previous inspections. • The relief valve is periodically removed and bench-tested by a certified individual. • The inlet and discharge piping are inspected to ensure that there is no blockage or corrosion that could impede proper functioning. • An internal inspection is performed to detect the onset of failure (such as corrosion, damaged internal components, or fouling/plugging). • The relief valve is returned to like-new condition prior to its return to service.
Basis for PFD and generic validation method
Consensus of the *Guidelines* subcommittee, based in general on *Guidelines for Pressure Relief and Effluent Handling Systems* (CCPS 1998b), Chapter 2, and recent published data (Bukowski and Goble 2009).

Dual Spring-Operated Pressure Relief Valves

Description: This IPL includes two spring-operated PRVs, each in service simultaneously with the full capacity necessary to individually prevent overpressure. There is a significant risk of common cause failure in a dual relief valve system. Both relief valves are exposed to the same process conditions, with a common risk of fouling or plugging. In addition, the valves may be of the same model and/or manufacturer and may receive maintenance at the same time by the same individual. As a result of the risk of common cause failure, the *Guidelines* committee did not suggest using a combined value for the dual PRV system, which would be equivalent to giving full credit for each of the two single PRVs (PFD_1 x PFD_2), or 0.0001. Instead, the generic PFD recommended in the data table is increased by a factor of 10, to a combined PFD of 0.001 for the dual relief valve system. To use a lower PFD, it is recommended that a thorough evaluation of the equipment, process conditions, and management systems be performed to ensure that the potential for common cause failure is sufficiently managed.

This PFD applies to a system that does not have valves that could isolate either relief device from the protected process. In the event that the closure of a single valve could isolate one relief device from the process, the suggested PFD of the combined system is 0.01. If the closure of a single valve could isolate both relief devices simultaneously, then the suggested PFD of the combined system is 0.1, equivalent to that of a single relief device that could be isolated from the process by one valve. (This does not apply to 3-way valves that cannot impede flow in any valve position.) Where robust management systems are in place to ensure that block valves upstream or downstream of relief devices remain in their correct positions, a lower PFD may be appropriate. Individual firms can claim a failure rate that reflects their specific system performance.

Obstruction of the relief path can also occur due to freezing. If freezing is possible, heat tracing of the piping and relief device is usually required. In this situation, the PFD of the dual relief device combination may be limited by the reliability of the heat tracing system. In freezing service, it is recommended that the analyst develop a PFD that is applicable to the specific system being analyzed.

Some installations provide two pressure relief valves at the same set pressure to allow one of the relief valves to serve as installed spare (back-up). For these installations, even if both pressure relief valves can be put in service simultaneously, the spare relief valve should not be claimed as an IPL due to the potential for chattering, which causes mechanical damage to the valve. The installation credited in this data table assumes that the dual pressure relief valves are always in service with staged set pressures in compliance with ASME Boiler and Pressure Vessel Code (BPVC) Section VIII (ASME 2013).

Consequences Avoided: Typically, the consequence being analyzed is a rupture of piping or a vessel. (Refer to Appendix E for additional information.)

Data Table 5.16. Dual spring-operated pressure relief valves

IPL description
Dual spring-operated pressure relief valves

Generic PFD suggested for use in LOPA
These values represent the PFDs of the combined dual-relief valve system.

No isolating valves present: 0.001
Single valve that could isolate one PRV: 0.01
Single valve that could isolate both PRVs simultaneously: 0.1

If a robust management system is in place to ensure that valves are returned to service in their proper positions after maintenance and remain in the appropriate state during operation, it may be appropriate to claim an improved PFD reflective of actual system performance. A better PFD may be applicable if a thorough review of the equipment, process conditions, and management systems indicates that the potential for common cause failure is sufficiently managed.

Special considerations for use of generic PFD for this IPL

- The PRVs are sized for the scenario being considered.
- The inlet and outlet piping are sized and mechanically designed properly.
- The relief valves are in clean service, and the metallurgy is corrosion-resistant to the particular service.
- If the service under evaluation has the potential for freezing of the process fluid before or during relief, development of a site-specific PFD is recommended.

Generic validation method

- The ITPM frequency is set in accordance with the manufacturer's recommendations and/or code requirements and may be adjusted based on the results of previous inspections.
- The relief valves are periodically bench-tested by a certified individual.
- The inlet and discharge piping are inspected to ensure that there is not blockage or corrosion that could impede proper functioning.
- An internal inspection is performed to detect the onset of failure (such as corrosion, damaged internal components, or fouling/plugging).
- The relief valves are returned to like-new condition prior to its return to service.

Basis for PFD and generic validation method
Consensus of the *Guidelines* subcommittee. The values of the PFDs are based in general on *Guidelines for Pressure Relief and Effluent Handling Systems* (CCPS 1998b), Chapter 2, and recent published data (Bukowski and Goble 2009).

Single Spring-Operated Pressure Relief Valve in Potential Plugging Service

No LOPA credit is given for a relief valve in fouling service, as it is assumed to be plugged and therefore unavailable to function properly on demand. The PFD could potentially be improved to match "clean service" relief valves with sufficient preventive measures. However, spring-operated pressure relief valves are generally not effective in services where plugging can be caused by polymerization during venting.

Pilot-Operated Pressure Relief Valve

Description: These valves are opened by the action of a small pilot device (usually a small spring-operated valve). These devices can be operated closer to the set pressure than conventional spring-opposed valves. However, the pilot tube that senses the pressure is usually of a small diameter, and this can increase the potential for blockage when in fouling or plugging service. In addition, if condensable gases can be present, a dual-phase pilot operated valve may be required to handle both liquids and gases.

Consequences Avoided: Typically, the consequence being analyzed is catastrophic rupture of piping or vessel. Generally, the onset of leakage of the piping or vessel is expected at pressures exceeding 121% of design. (Refer to Appendix E Considerations for Overpressure and Related Outcomes.)

Figure 5.22. Pilot-operated pressure relief valve, courtesy of Emerson Process Management.

Data Table 5.17. Pilot-operated pressure relief valve

IPL description
Pilot-operated pressure relief valve
Generic PFD suggested for use in LOPA
0.01 for failure to open enough at set pressure (100% of rating) If there is an isolation valve (block valve) upstream or downstream of the relief device or on the pilot line, then the suggested PFD is 0.1, unless there is a management system in place to ensure that valves are returned to service in their proper positions after maintenance and that they remain in the appropriate state during operation.
Special considerations for use of generic PFD for this IPL
• The PRV is sized for the scenario being considered. • The inlet and outlet piping are sized correctly and are mechanically adequate for relief flow. • The relief valve is in clean service, and the metallurgy is corrosion-resistant to the particular service. • The service under evaluation does not have the potential for freezing of the process fluid before or during relief; if freezing is possible, then adequate heat tracing of the relief valve and piping is installed and maintained.
Generic validation method
• The ITPM frequency is set in accordance with the manufacturer's recommendations and/or code requirements and may be adjusted based on the results of previous inspections. • The relief valve is periodically removed and bench-tested by a certified individual. • The inlet and discharge piping are inspected to ensure that there is not blockage or corrosion that could impede proper functioning. • An internal inspection is performed to detect the onset of failure (such as corrosion, damaged internal components, or fouling/plugging). • The relief valve is returned to like-new condition prior to its return to service.
Basis for PFD and generic validation method
Consensus of the *Guidelines* subcommittee that the PFD is comparable to that of spring-operated PRVs.

Gas Balance/Adjustable Set Pressure Surge Relief Valve

Description: The fundamental requirements of a surge relief system include the need for a high-capacity valve that can open very quickly to remove surge pressures from the line and then return to the normal (closed) state quickly without causing additional pressure surge during closure. A surge relief valve (SRV) is often required to open fully in a very short period of time so that the entire flowing stream may be relieved, if necessary.

Consequences Avoided: A properly sized surge relief valve can prevent overpressurization of a tank or vessel as a result of high pressure surges.

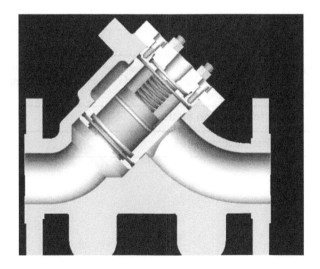

Figure 5.23. Surge relief valve, courtesy of Emerson Process Management.

Data Table 5.18. Gas balance/adjustable set pressure surge relief valve

IPL description
Gas balance/adjustable set pressure surge relief valve
Generic PFD suggested for use in LOPA
0.01 If there is an isolation valve (block valve) upstream or downstream of the relief device, then the suggested PFD is 0.1, unless there is a management system to ensure that valves are returned to service in their proper positions after maintenance and that they remain in the appropriate state during operation.
Special considerations for use of generic PFD for this IPL
• Valves may or may not include a set pressure spring. • The relief valve is in clean service, and the metallurgy is corrosion-resistant to the particular service. • The surge relief valve is sized for the scenario being considered. • The inlet and outlet piping are sized correctly and are mechanically adequate for relief flow. • Modeling confirms that response time is adequate for the application.
Generic validation method
Inspection for plugging or fouling of the overflow line is performed based on the frequency suggested by the manufacturer and/or code requirements and may be adjusted based on previous maintenance history.
Basis for PFD and generic validation method
Consensus of the *Guidelines* subcommittee.

Buckling Pin Relief Valve

Description: The buckling pin relief valve (BPRV), also referred to as a "rupture pin," is a relief device that is similar to a spring-loaded relief valve except that it does not reclose on its own. The relief valve opens when the pin that holds the valve closed is bent due to the force of the increased pressure in the protected system. Bending of the pin is similar to overloading the spring on a spring-loaded relief valve. Buckling pins are approved in the ASME Boiler and Pressure Vessel Code (BPVC) Section VIII (ASME 2013), and they are effective for overpressure protection of a pressure vessel.

Buckling pins are easier to install than many other types of relief devices and more difficult to install incorrectly. BPRVs are also of a simpler design than spring-loaded or pilot-operated relief valves. There are no specific reliability data available on buckling pin relief devices, but the failure rates for ASME BPVC Section VIII (2013) pressure relief devices are assumed to be representative of a buckling pin device. The generic value provided in the data table is applicable for clean, noncorrosive, nonfouling service.

Consequences Avoided: Proper operation of a buckling pin relief device will prevent overpressurization of a vessel or other protected equipment. Typically, the consequence being analyzed is catastrophic rupture of piping or vessel. Generally, the onset of leakage is expected at pressures exceeding 121% of design. (Refer to Appendix E Considerations for Overpressure and Related Outcomes.)

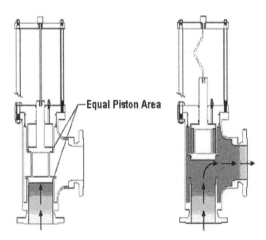

Figure 5.24. Buckling pin relief valve, courtesy of Rupture Pin Technology.

Data Table 5.19. Buckling pin relief valve

IPL description
Buckling pin relief valve
Generic PFD suggested for use in LOPA
0.01 for failure to open enough at set pressure (100% of rating), if properly inspected If there is an isolation valve (block valve) upstream or downstream of the relief device, then the suggested PFD is 0.1, unless there is a management system in place to ensure that valves are returned to service in their proper positions after maintenance, and that they remain in the appropriate state during operation.
Special considerations for use of generic PFD for this IPL
• The buckling pin relief device is confirmed to be sized for the scenario being considered. • The inlet and outlet piping are sized correctly and mechanically adequate for relief flow. • The relief valve is in clean service, and the metallurgy is corrosion-resistant to the particular service. • The service does not have the potential for freezing of the process fluid before or during relief; if freezing is possible, then adequate heat tracing of the relief device and piping is installed and maintained.
Generic validation method
The inspection frequency is based on company experience and the results of previous inspections. • The *Guidelines* subcommittee recommends initial inspection within 1 year of installation. • Inspection of the buckling pin device is performed to ensure proper travel and proper setting/design limit of buckling pin. • Inspection includes the inlet and discharge piping to ensure that there is no fouling or corrosion that would impede proper functioning. • Internal inspection is performed to detect the onset of failure (corrosion, damaged internals, fouling/plugging). • The buckling pin is replaced prior to return to service.
Basis for PFD and generic validation method
Consensus of the *Guidelines* subcommittee. There are no specific reliability data available on buckling pin relief devices, but the failure rates for other pressure relief devices are assumed to be representative of a buckling pin device.

Buckling Pin Isolation Valve

Description: A buckling pin isolation valve (BPIV) is similar in concept to a BPRV, except that instead of venting the contents to relieve pressure, the overpressure causes an isolation valve to close and isolate the protected system from the source of the pressure. BPIVs are useful as emergency shutdown valves. The PFD of a BPIV is considered to be similar to that of a spring-operated relief valve when in clean (noncorrosive, nonfouling) service.

Consequences Avoided: Proper operation of a buckling pin isolation valve can prevent overpressurization of a vessel or system by stopping flow from a higher pressure source.

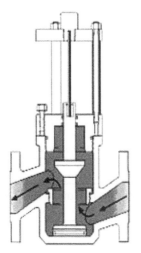 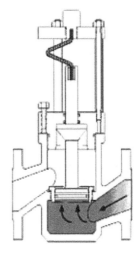

Figure 5.25. Buckling pin isolation valve, courtesy of Rupture Pin Technology.

Data Table 5.20. Buckling pin isolation valve

IPL description
Buckling pin isolation valve (BPIV)
Generic PFD suggested for use in LOPA
0.01 for failure to close at set pressure, if properly inspected If there is an isolation valve (block valve) upstream or downstream of the BPIV, then the PFD is taken as 0.1, unless there is a management system in place to ensure that valves are returned to service in their proper positions after maintenance, and that they remain in the appropriate state during operation.
Special consideration for use of generic PFD for this IPL
• The buckling pin valve is used in a clean service, and the metallurgy is corrosion-resistant to the particular service. • The buckling pin valve is properly rated to close at a pressure that will protect the downstream vessel.
Generic validation method
The inspection frequency is based on company experience and the results of previous inspections. • The *Guidelines* subcommittee recommends initial inspection within 1 year of installation. • Inspection of the buckling pin device is performed to ensure proper travel and proper setting/design limit of buckling pin. • Internal inspection of valve is conducted to detect any build-up that would limit complete valve closure. • Inspection is performed to check for any evidence of corrosion that would impede proper functioning. • The BPIV is returned to like-new condition prior to return to service.
Basis for PFD and generic validation method
Consensus of the *Guidelines* subcommittee. The failure rate of the BPIV expected to be similar to that of a BPRV.

Rupture Disk

Description: A rupture disk (RD) consists of a disk made of a variety of materials (i.e., graphite, ceramic, stainless steel, etc.). The design of the disk is such that it will rupture within a specified pressure range. These devices are used alone or in combination with relief valves. A rupture disk is usually employed when

- The process fluid has properties (polymerization, corrosion, fouling, etc.) that make the use of a relief valve undesirable, or similarly make the use of a relief valve alone undesirable

- The relief capacity needed is so great that use of relief valve is not feasible

The major disadvantage of rupture disks is that they cannot reclose; when the disk is burst, a large quantity of material can be discharged. Also, the rupture disk may burst due to a temporary spike in system pressure; this results in the need to replace the disk.

Consequences Avoided: A rupture disk can prevent overpressure, leading to rupture of piping or a vessel. Generally, the onset of leakage is expected at pressures exceeding 121% of design. (Refer to Appendix E Considerations for Overpressure and Related Outcomes for additional discussion.)

Figure 5.26. Intact and burst rupture disks, courtesy of CMC Technologies Pty. Limited/Marston Division of Safety Systems UK.

Data Table 5.21. Rupture disk

IPL description
Rupture disk
Generic PFD suggested for use in LOPA
0.01 for failure to open at burst pressure If there is an isolation valve (block valve) upstream or downstream of the relief device, then the PFD is taken as 0.1, unless there is a management system in place to ensure that valves are returned to service in their proper positions after maintenance and that they remain in the appropriate state during operation.
Special considerations for use of generic PFD for this IPL
• The rupture disk (RD) is confirmed to be sized for the scenario being considered. • The inlet and outlet piping are sized correctly and are mechanically adequate for relief flow. • The RD is in clean service, and the metallurgy is corrosion-resistant to the particular service. • The service under evaluation does not have the potential for freezing of the process fluid before or during relief; if freezing is possible, then adequate heat tracing of the RD and piping are installed and maintained.
Generic validation method
• Inspection frequency for incipient conditions is set as a function of the severity of the service and previous inspection history. • A visual check of stamp or tag on the RD will confirm the specification of the device. • A visual inspection of the inlet and outlet is done to detect incipient conditions, such as fouling or corrosion, that may cause the RD to burst at a higher pressure. • Inspection is typically done in situ, and can be by boroscopic inspection if a complete view of the surface is possible. • If the disk is not in a pre-torque holder or cannot be inspected in situ, the inspection requires disk replacement. • Disks are replaced on an appropriate frequency based on the service, past history, when they cannot be inspected in situ, or whenever the burst pressure may have been exceeded.
Basis for PFD and generic validation method
Consensus of the *Guidelines* subcommittee. See *Guidelines for Pressure Relief and Effluent Handling Systems* (CCPS 1998b), Chapter 2, for guidance.

Spring-Operated Pressure Relief Valve with Rupture Disk

Description: This combination of devices consists of a relief valve with an upstream rupture disk. Rupture disks are commonly employed in tandem with relief valves; rupture disks can protect the relief valve from corrosive, fouling, or plugging services that would make use a relief valve alone undesirable. It is important that the rupture disk be of a nonfragmenting variety to prevent blockage and/or reclosing of the relief valve. In potential plugging service, the process side of the rupture disk may need to be periodically flushed to keep it clean. The major disadvantage of rupture disks is that they cannot reclose; using the PRV/rupture disk combination addresses this issue. However, spring-operated pressure relief valves are generally not effective in services where plugging can be caused by polymerization during venting.

> WARNING: If the rupture disk has a pin-hole leak, then the pressure can be equalized on both sides of the disk. Should a high pressure event occur, the rupture disk may not relieve when intended, and the PRV/RD combination may be ineffective. Pressure monitoring between the rupture disk and the relief valve allows for early detection and correction of a leak in the rupture disk.

Consequences Avoided: Typically, the consequence being avoided is overpressure, leading to the rupture of piping or a vessel. Generally, the onset of leakage is expected at pressures exceeding 121% of design. (Refer to Appendix E Considerations for Overpressure and Related Outcomes for additional discussion.)

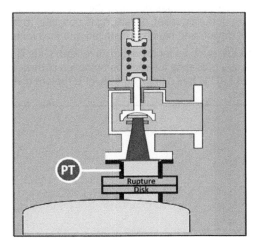

Figure 5.27. Spring-operated relief valve with rupture disk.

Data Table 5.22. Spring-operated pressure relief valve with rupture disk

IPL description
Spring-operated pressure relief valve with rupture disk
Generic PFD suggested for use in LOPA
0.01 for failure to open enough at set pressure If there is an isolation valve (block valve) upstream or downstream of the relief device, then the PFD is taken as 0.1, unless there is a management system in place to ensure that valves are returned to service in their proper positions after maintenance and that they remain in the appropriate state during operation.
Special considerations for use of generic PFD for this IPL
The rupture disk and relief valve meet the special considerations for the individual devices; refer to Data Tables 5.15 and 5.21. • The downstream side of the rupture disk is monitored to ensure the continued integrity of the RD. • Monitoring of the space between the RD and PRV can be done using a pressure transmitter or switch with alarm or a pressure gauge with routine operator inspections. Use of the PRV/RD combination in plugging service may also require flushing of the process side of the rupture disk to keep it clean. • A nonfragmenting type of rupture disk is used to avoid blockage and/or reclosing of the relief valve by debris. • The metallurgy is corrosion-resistant to the particular service. • The service under evaluation does not have the potential for freezing of the process fluid before or during relief; if freezing is possible, then adequate heat tracing of the relief device and piping is installed and maintained. Certain industry groups, such as those within coverage of the Chlorine Institute, can follow their industry guidance for such combination of devices. For instance, the Chlorine Institute (2011) has design criteria for a rupture disk that is integral to the relief valve body.

Data Table 5.22. Continued

Generic validation method
<ul><li>The inspection frequency for incipient conditions is set based on the severity of the service and previous inspection history.</li><li>A visual check of stamp or tag on the RD confirms the specification of the device.</li><li>Inspections include the inlet and discharge pipe and disk.</li><li>If the disk is not in a pre-torque holder or cannot be inspected in situ, the inspection requires disk replacement.</li><li>The inspector should record the pressure reading between the RD and PRV and validate that ongoing pressure monitoring is being done.</li><li>The relief valve is periodically removed and bench tested by a certified individual.</li><li>The inlet and discharge piping are inspected to ensure that there are not blockage or corrosion that could impede proper functioning.</li><li>An internal inspection will allow detection of the onset of failure (such as corrosion, damaged internal components, or fouling/plugging).</li><li>Disks are replaced on an appropriate frequency based on the service, past history, when they cannot be inspected in situ, or whenever the burst pressure may have been exceeded.</li><li>The relief valve is returned to like-new condition prior to its return to service.</li></ul>
Basis for PFD and generic validation method
Consensus of the *Guidelines* subcommittee. See *Guidelines for Pressure Relief and Effluent Handling Systems* (CCPS 1998b), Chapter 2, for guidance.

Conservation Vacuum and/or Pressure Relief Vent

Description: Conservation vacuum and/or pressure relief vents (VPRVs) are also known as conservation vents or pressure-vacuum vent valves. These devices are used to prevent overpressure (such as by venting displacing vapor volume when filling a tank), and/or to prevent a vacuum (such as by introducing air when liquid is being withdrawn from a tank). They also minimize vapor emissions from the tank and are often used on tanks that contain liquids with low flash points. Conservation vents are also found in applications where an inert or pad gas is used on the vapor space of the tank.

Conservation vents are typically used on atmospheric tanks, low pressure vessels, or containers. A conservation vent may be spring-loaded or weight-loaded and is generally mounted to the vapor space of the tank using a flange.

When evaluating a conservation vent as an IPL, it is important to consider the mode of operation of the device. If a vessel has a primary means of relieving vacuum (such as a pad gas) or relieving pressure (such as a vent to an abatement system), then a conservation vent may appropriately be treated as an IPL in LOPA. If, however, the conservation vent is the vessel's primary means of relief, the conservation vent would be expected to operate frequently. The conservation vent would be in high demand mode, and failure of the conservation vent would be treated as an initiating event rather than an IPL. Refer to Section 3.5.2 for a discussion of high demand mode.

It is important to periodically inspect conservation vents for blockage as a result of snow, ice, nests, or other debris. Conservation vents may be supplied with a flame arrester to prevent external flame from entering a flammable tank headspace. If a flame arrester is installed in conjunction with the conservation relief vent, the flame arrester is assumed to be in clean service and not subject to fouling from process vapors. Flame arresters are also periodically inspected for plugging. (Refer to the discussion of flame arresters in Section 5.2.1.)

Consequences Avoided: Proper operation of a conservation relief vent will prevent overpressure and/or vacuum creation in the tank. Vacuum is less likely than overpressure to generate a significant leak; tank damage and minor leakage is more common.

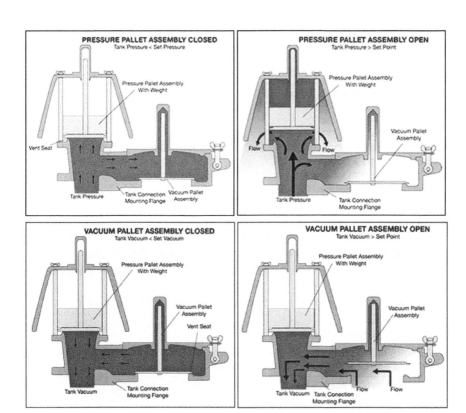

Figure 5.28. Conservation pressure/vacuum relief vent, courtesy of the Protectoseal Company.

Data Table 5.23. Conservation vacuum and/or pressure relief vent

IPL description
Conservation vacuum and/or pressure relief vent
Generic PFD suggested for use in LOPA
0.01 If there is an isolation valve (block valve) upstream or downstream of the relief device, then the PFD is taken as 0.1, unless there is a management system in place to ensure that valves are returned to service in their proper positions after maintenance and that they remain in the appropriate state during operation.
Special consideration for use of generic PFD for this IPL
• The relief vent is confirmed to be sized for the scenario being considered. • The vent is properly implemented; one reference is API 2000 (2009c). • The inlet and outlet piping are sized correctly and are mechanically adequate for venting. • The vent is in clean service, and the metallurgy is corrosion-resistant to the particular service. • The device is operating in low demand mode. • If freezing of the process fluid or atmospheric moisture is possible, then adequate heat tracing of the vent and piping is installed and maintained.
Generic validation method:
• The inspection frequency is based on the manufacturer's recommendation and may be adjusted based on previous inspection data. • The inspection frequency is based on similar service, related codes/standards, and actual performance data. • A visual inspection is performed to identify signs of incipient failure, and the device is restored to as-new condition prior to return to service.
Basis for PFD and generic validation method:
Consensus of the *Guidelines* subcommittee. No specific failure data are available on spring-loaded or weight-loaded conservation vacuum and/or pressure relief vents. Failure rates for ASME Section VIII (2013) pressure relief valves are assumed to be representative of spring-loaded and weight-loaded conservation vents. See API 2000 (2009c) and API 221 (1991) for guidance on installation and validation.

Vacuum Breaker

Description: A vacuum breaker, also known as a vacuum relief valve (VRV) or a vacuum safety valve (VSV), is a device placed on an atmospheric or low pressure-rated vessel to prevent vacuum-induced collapse.

A similar concept can also be used in a fluid line to prevent backflow in lieu of a barometric leg. A vacuum breaker in a fluid line typically contains a diaphragm disk that is pressed forward by supply pressure to cover small vent holes. Should the supply pressure drop, the disk springs back, opening the vent holes and allowing outside air to enter.

As with VPRVs, it is important to consider the mode of operation of the device. If a vessel has a primary means of relieving vacuum (such as a pad gas), then a vacuum breaker may appropriately be treated as an IPL in LOPA. If, however, the vacuum breaker is the vessel's primary means of vacuum relief, the vacuum breaker would be expected to operate frequently. The vacuum breaker would then be in high demand mode, and failure of the vacuum breaker would be treated as an initiating event rather than an IPL. Refer to Section 3.5.2 for a discussion of high demand mode.

Consequences Avoided: Vacuum breakers prevent collapse and damage to a vessel. Vacuum breakers also prevent siphoning back into the protected system; this helps to prevent contamination if the system pressure drops. Vacuum is less likely than overpressure to generate a significant leak; tank damage and minor leakage is more common.

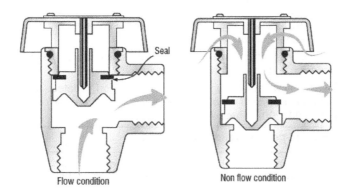

Figure 5.29. Atmospheric vacuum breaker (U.S. EPA 2003).

Data Table 5.24. Vacuum breaker

IPL description
Vacuum breaker
Generic PFD suggested for use in LOPA
0.01
If there is an isolation valve (block valve) upstream or downstream of the vacuum breaker, then the PFD is taken as 0.1, unless there is a management system in place to ensure that valves are returned to service in their proper positions after maintenance and that they remain in the appropriate state during operation.
Special considerations for use of generic PFD for this IPL
• The vacuum safety valve, vacuum relief valve, or vacuum breaker is confirmed to be sized for the scenario being considered.
• The inlet and outlet piping are sized correctly and are mechanically adequate for vacuum relief flow.
• The service is considered nonfouling, and the vacuum breaker has no history of fouling.
• The metallurgy is corrosion-resistant to the particular service.
• The device is operating in low demand mode.
• The service does not have the potential for freezing of the process fluid before or during relief. If freezing is possible, then adequate heat tracing of the vacuum relief valve and piping is installed and maintained
Generic validation method
The inspection frequency is based on previous inspection data and the manufacturer's recommendation.
• The site sets an inspection frequency based on similar service, related codes/standards, and actual performance data.
• A visual inspection of the vacuum breaker is performed to detect incipient signs of failure, and the device is restored to as-new condition prior to return to service. API 2000 (2009c) provides useful guidance.
Basis for PFD and generic validation method
Consensus of the *Guidelines* subcommittee. See *Guidelines for Safe and Reliable Instrumented Protective Systems* (CCPS 2007b), p 288, Table B.4.

Frangible Roof on Flat-Bottom Tank

Description: Atmospheric and low-pressure aboveground flammable liquid storage tanks require emergency venting that is designed to relieve excessive pressure caused by exposure to fire. One means to meet this requirement is a roof-to-shell connection that will fail before buoyant forces raise the tank (round the bottom), which could ultimately stress the bottom-to-shell joint, causing it to fail. This type of design is referred to as a "frangible roof." The purpose of a frangible roof is to provide a large venting area to protect against a bottom-seam failure of a flat-bottom storage tank. Frangible roofs are useful in providing emergency venting when other means of venting are impractical.

The codes used to design aboveground flammable liquid storage tanks give various methods for designing the roof-to-shell connection and anchors to hold the shell flat against the pad. These codes include API-650 (2013), API-2000 (2009c), NFPA-30 (2008a), US OSHA 29 CFR 1910.106 (2005), and UL-142 (2007).

Consequences Avoided: The outcome avoided by the use of a frangible roof is the failure of the bottom-to-shell joint, which could result in consequences such as the tank being propelled from its foundation and instantaneously releasing the contents of the tank. If a frangible roof does open, however, there is an alternative consequence to be considered. The tank bottom and shell remain intact, but the top seam of the tank fails. This could still result in a vapor cloud release or a fireball. Therefore, whether or not a frangible roof is considered to be a valid IPL depends on whether or not it prevents the specific consequence of concern. A frangible roof may limit damage to the environment and nearby process equipment but may not sufficiently protect humans from a vapor cloud or fireball hazard. When a frangible roof is used as an IPL to prevent the failure of the bottom-to-shell joint, the alternative scenario of the failure of the top seam of the tank should also be assessed.

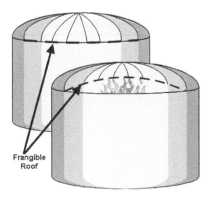

Figure 5.30. Frangible roof on a flat-bottom tank.

Data Table 5.25. Frangible roof on flat-bottom tank

IPL description
Frangible roof on flat-bottom tank
Generic PFD suggested for use in LOPA
0.01
Special considerations for use of generic PFD for this IPL
The purpose of a frangible roof is to provide venting area to protect against a bottom-seam failure of a flat-bottom storage tank, resulting in the tank being propelled from its foundation and instantaneously releasing the tank contents. Frangible roofs are useful in providing emergency venting where such a large venting area is required that other means of venting are impractical. The tank design basis is reviewed to ensure that the tank is designed appropriately and well-secured to the foundation. Key design features include: • The top seam will fail at the target pressure and below the lift pressure of the tank. • The specifications of the top angle ring (or its supports) are not exceeded since this is the critical element that is expected to fail first. Also, anything that increases the strength/stiffness of the top head (beyond specifications) could increase the failure pressure and is evaluated to ensure that the design will perform as originally specified. • The design of the tank anchor chairs, bottom ring, number of anchor bolts, and the diameter and length of the bolts is sufficient to keep the tank in place during a relief event. • The tank foundation is in an acceptable condition, with the anchor bolts well-secured to the foundation. There are no walkways, piping, or other obstructions that would prevent the frangible roof from fully opening and relieving the necessary pressure.

Data Table 5.25. Continued

Generic validation method
• Periodic visual inspections are done of the top weld, roof, and top anchor ring for signs of corrosion or cracking.
• Visual inspection of bottom chairs, bolts, and concrete are also done for signs of corrosion or cracking to ensure that the tank would remain securely anchored in the event of a frangible roof release.
• For external inspections, nonintrusive ultrasonic measurements are taken by a competent inspector of the shell thickness, roof thickness, and nozzles, and the measurements are compared to the original design drawings. API 653 (2009a) is one reference for developing an inspection protocol.
• External visual and nonintrusive ultrasonic measurements are done by a competent tank inspector every five years (or more frequently, as determined by regulations or based on the findings of previous inspections).
• Internal inspections are performed by a competent tank inspector/engineer at a minimum frequency of every 15 years (or more frequently as determined by regulations or based on service conditions and the findings of previous inspections).
• The results of visual inspections, ultrasonic measurements, and internal inspections are documented.
Basis for PFD and generic validation method
Consensus of the *Guidelines* subcommittee. References for design include: API Standard 650 *Welded Steel Tanks for Oil Storage,* 12th Edition (2013); API Standard 653 *Tank Inspection, Repair, Alteration, and Reconstruction* (2009a); and API Publication 937 *Evaluation of the Design Criteria for Storage Tanks with Frangible Roofs Joints* (1996).

Explosion Isolation and Venting

Chemical processing systems usually consist of several pieces of equipment linked by pipelines. An explosion in one piece of equipment can propagate through these pipelines to other connected equipment. Isolation is a method that can be used to prevent flame propagation and exposure of the connected vessel to a pressure wave.

Active isolation systems utilize a signal from a pressure, flame, or rupture panel burst detector to trigger the operation of one or more isolation devices. Passive isolation systems consist of mechanical valves that use the pressure wave from the explosion to close the isolation valves. When activated, the isolation devices close rapidly to prevent the flame front and associated pressure wave from traveling to other connected equipment.

NFPA 654 (2013) provides guidance on when isolation devices should be provided to prevent explosion propagation between connected equipment. Also refer to NFPA 69 (2008c) for additional guidance regarding explosion isolation valves and their use.

Deflagration venting is another technique for explosion protection. It is widely used due to its relative simplicity, effectiveness, and cost versus other types of systems. Deflagration vents can be used to protect both process equipment and structures. The vents open at a specified pressure to permit the release of combustion gases and burning solids rapidly enough to prevent an overpressure of the process equipment or building.

Vents are designed to discharge to a safe location. A vent duct is designed to be as short as possible and generally has a cross section as least as great as the vent to ensure sufficient venting area. Duct bends are minimized since they can increase pressure during venting. When a vent cannot be directed outside of a building or to a safe location, a flameless venting device may be used. As gases are vented from a flameless venting device, dust is captured, the flame front is cooled, and no flame is released to the environment.

Another available means of explosion protection is a deflagration suppression system. This system includes a sensor with sufficient sensitivity to detect an incipient explosion. Upon detection, a fast-acting valve rapidly introduces a suppressant into the equipment to extinguish the flame before overpressure develops. These sophisticated systems are designed for the specific application, and a generic PFD could not be developed for a deflagration suppression system.

The subsequent data tables cover three typical explosion isolation/venting applications:

Data Table 5.26. Explosion isolation valve
Data Table 5.27. Explosion panels on process equipment
Data Table 5.28. Vent panels on enclosures

Explosion Isolation Valve

Description: The explosion isolation valve is a fast-acting mechanical barrier that blocks a potential flame path leading to other process equipment. The valve remains open until it receives a signal from a pressure responder and/or an optical sensor that detects hot particles, embers, and flames. Upon detection, actuation of the gas cartridge is initiated to rapidly isolate the pipe. The valve responds quickly, closing in milliseconds. This quick reaction prevents the passage of hot particles, embers, flames, or pressure from continuing to flow through the pipeline. However, it does not prevent or mitigate the consequences of the event in the section of the process where the deflagration started.

NFPA 654 (2013) provides guidance on when isolation devices should be provided to prevent explosion propagation between connected equipment. Also refer to NFPA 69 (2008c) for additional guidance regarding explosion isolation valves and their use.

Consequences Avoided: The explosion isolation valve protects against the propagation of flame between interconnected equipment.

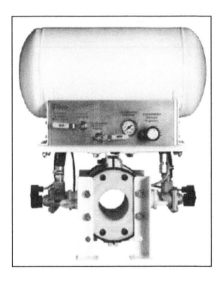

Figure 5.31. Explosion isolation valve, courtesy of Fike Corporation.

Data Table 5.26. Explosion isolation valve

IPL description
Explosion isolation valve
Generic PFD suggested for use in LOPA
0.1
Special consideration for use of generic PFD for this IPL
Refer to NFPA 69 *Standard on Explosion Prevention Systems* (2008c). Explosion isolation valves can mitigate deflagration scenarios but not detonation scenarios.
Generic validation method
• The inspection frequency is based on the manufacturer's recommendations and the results of previous inspections. • Inspection and testing can be performed in situ or off-line and include an internal inspection for conditions that could indicate the onset of failure (such as fouling, plugging, sticking, and corrosion). Refer to NFPA 69 *Standard on Explosion Prevention Systems* (2008c) for details.
Basis for PFD and generic validation method
Consensus of the *Guidelines* subcommittee. Refer to NFPA 69 *Standard of Explosion Prevention Systems* (2008c). NFPA 654 (2013) provides guidance for solids handling systems.

Explosion Panels on Process Equipment

Description: Explosion panels work by sacrificing a portion of the equipment enclosure (analogous to bursting of a rupture disk) if an internal deflagration occurs that rapidly increases the pressure inside the enclosure. Explosion panels are designed to protect against vessel or duct overpressure due to an internal dust/vapor/gas deflagration. These are properly designed per NFPA 68 (2007) or other appropriate standard based on the specific material that is being handled in the equipment.

The explosion is vented to a safe atmospheric location, and the safe location is maintained and controlled as a safe location for discharge. Failing to meet this requirement can prevent an explosion panel from serving as an IPL for human harm scenarios. See NFPA 68 (2007) or other appropriate standard for methods for predicting fireball size to determine safe distances from explosion vents. The fireball size is often much larger when venting dust explosions than when venting vapor/gas explosions. Designs are available that can quench the vented discharge flame and significantly reduce the potential effects outside of the process equipment or enclosure.

Consequences Avoided: Proper operation of explosion panels during an internal dust/vapor/gas explosion can protect a vessel or duct from excessive overpressure; however, to protect from human harm, the expelled flame and hot products of combustion are routed to a safe location. Also, the explosion panel is designed or restrained in such a way as to not pose a projectile hazard when the explosion vent opens.

Figure 5.32. Vent panel and flameless vent, courtesy of Fike Corporation.

Data Table 5.27. Explosion panels on process equipment

IPL description
Explosion panels on process equipment
Generic PFD suggested for use in LOPA
0.01
Special considerations for use of generic PFD for this IPL
• Explosion panels designed to protect against vessel overpressure due to a dust/vapor/gas explosion are properly designed per NFPA 68 (2007) or other appropriate standard for the material being handled in the equipment. • Associated piping systems have proper relief design or have appropriate explosion prevention systems designed as per appropriate standards. • The explosion is routed to a safe location. • Explosion panels can protect against internal deflagrations but not detonations.
Generic validation method
• The inspection frequency is determined based on the manufacturer's recommendations and the results of previous inspections. • Inspection of the panel is done to assure that the panel is properly installed and that there is no buildup on the panels.
Basis for PFD and generic validation method
Consensus of the *Guidelines* subcommittee. No specific failure data are available for the PFD of explosion panels, but the technology is similar to rupture disks. Therefore, the entry has been given the same PFD. Refer to NFPA 68 (2007), NFPA 69 (2008c), EN 14491 (BS 2006), or other appropriate standards for guidance on design.

Vent Panels on Enclosures

Description: Vent panels or doors open to provide relieve for an enclosure or room from the hot, expanding deflagration gases. Deflagration vents open at a predetermined pressure referred to as P_{stat}. The pressurized gases are discharged to the atmosphere either directly or via a vent duct, resulting in a reduced deflagration pressure, P_{red}. The deflagration vent arrangement is designed to ensure that pressure, P_{red}, is below the rupture pressure of the enclosure or room.

When designing vent closure devices, it is important that the vent open at the lowest practical pressure while avoiding inadvertent opening and leakage of material (Grossel and Zalosh 2005). The panel is designed to be resistant to the materials being handled and strong enough to withstand normal operating pressures. The vent panel needs to open quickly without becoming a hazardous projectile.

Building or room vent closures can have a variety of designs:

- Hinged doors, windows, and panel closures
- Shear and pull-through fasteners
- Friction-held closures
- Weak roof or weak wall construction
- Large-area panels

Additional discussion can be found in NFPA 68 (2007).

Consequences Avoided: Vent panels prevent damage to an enclosure or room. However, activation of the panel does result in a pressure wave and loss of containment of vapor/gas. If the vent panel relieves into an occupied area, a vent panel may not be an effective IPL against impact to nearby workers. However, a vent panel may be an effective IPL against human harm if the release vents to an unoccupied area. The impact of vent panel activation needs to be clearly understood to determine if vent panels will protect against a specific scenario.

Figure 5.33. Example of a vent panel on an enclosure, courtesy of Haz-Safe L.L.C.

Data Table 5.28. Vent panels on enclosures

IPL description
Vent panels on enclosures
Generic PFD suggested for use in LOPA
0.01
Special considerations for use of generic PFD for this IPL
Vent panels are designed: • With sufficient relief area to prevent overpressure at the maximum temperature expected. • By a qualified expert using a proven method such that the panels will operate with a fast enough response time to effectively mitigate the event. This may require consideration of the momentum and backpressure of panels during an event as well as the dynamic loading of receptors (e.g., walls). • For potential vacuum in the equipment, if applicable, such that the panels will remain reliable under actual process conditions. • To incorporate means to prevent injury to personnel and other equipment. These may include hinges and cables to prevent the panels from becoming projectiles and restriction of personnel entry into the venting area. • Such that ductwork has a clear path, with no resistance to flow that can impede the effectiveness of the device. • Designed such that the panels can be periodically inspected/maintained. References for design include NFPA 68 (2007), European VDI 3673 (2002), and ATEX Directive 94/9/EC (2009).
Generic validation method
• The inspection frequency varies depending on the service and based on the manufacturer's recommendation. • The vent panel is included on a routine maintenance schedule, including inspection for any fouling or corrosion that might affect the available relief area or the ability of the relief to function with sufficient speed. • Hinges or other devices used to secure the vent panel are inspected for damage or metal fatigue. • Gaskets are replaced as specified by the manufacturer. • For dust applications where equipment changes might affect the need for a larger relief area, the application is included in an MOC process.
Basis for PFD and generic validation method
Consensus of the *Guidelines* subcommittee based on consistency with the PFD for a rupture disk.

5.2.2.3   Other Mechanical Devices

This section discusses the many types of self-contained mechanical devices that can be valid IPLs. Mechanical devices can be driven by process-variable forces and can behave as either analog devices (such as a pressure regulator) or as discrete action devices (such as an excess flow valve). This section lists IPLs that are mechanical devices that have not been discussed elsewhere in this document.

Excess Flow Valve

Description: Excess flow valves are mechanical devices designed to stop the flow of a material when a predetermined flow rate or differential pressure is reached. A typical excess flow valve is self-contained, with the entire mechanism contained inside of the valve body.

There are many designs of excess flow valves. Valve body sizes ranging from tubing (1/2 inch) to large piping system (10 inch or larger) are widely available. Typical designs include

- Spring-loaded
- Magnetic release ring
- Differential pressure Venturi effect piston closure
- Weighted ball

Excess flow valves are sometimes installed on the bottom outlets of anhydrous ammonia storage tanks and liquefied petroleum gas (LPG) storage tanks. Internal excess flow valves can be installed on the inside of chlorine and anhydrous ammonia rail cars. These installations are intended to prevent the release of the tank contents given a catastrophic line failure at the tank or very near to the tank. Excess flow valves can also be installed in-line in process piping.

An excess flow valve is not an IPL that prevents loss of containment; rather, it acts to limit loss of containment by preventing a larger release. Whether or not an excess flow valve is a valid IPL for a scenario depends on the consequence of concern. If a significant consequence occurs as a result of a small spill, an excess flow valve may not be a valid IPL since some release will occur even upon successful activation. Further, the excess flow valve is typically not designed to protect against certain types of leak scenarios as described below.

Whichever physical mechanism of closure is used, valve closure ultimately relies upon the forces generated in the valve housing as the fluid flows through it. Unless the fluid flow rate exceeds the valve design closure rate, the valve will not close. A number of cases have occurred when the valve failed to close in the event of a small leak, flange gasket failure, or small diameter line failure at a distance from the excess flow valve (see alerts by US EPA [2007] and US CSB [2007]). Calculation methods are available (Freeman and Shaw 1988) that allow for the estimation of the flow rate that can be expected at the excess flow valve given the piping geometry, fluid properties, and release scenario. If the calculation indicates that the fluid flow will exceed the design closure flow rate of the valve, then the excess flow valve is potentially effective for stopping the release. In summary, if the consequence of concern is a large release at high flow, such as a full-bore rupture, then an excess flow valve may be considered as an IPL if properly specified, installed, and maintained (including periodic inspection or replacement).

Excess Flow Valve

Description: Excess flow valves are mechanical devices designed to stop the flow of a material when a predetermined flow rate is reached. They are mitigative IPLs intended to prevent catastrophic loss of vessel contents in the event of a line rupture. Failure mechanisms of concern include blockage of the valve, corrosion of the valve seats, mechanical damage to the valve internals, and vibration-induced leakage or valve failure.

Excess flow valves have extensive design considerations and limitations that should be considered during LOPA evaluation. Excess flow valves do not protect against scenarios where the consequence of concern is a leak, and additional LOPA scenarios are needed for those cases where the flow rate is not high enough to close the excess flow valve. Calculation methods are available (Freeman and Shaw 1988) that allow for the estimation of the flow rate required to exceed the design flow rate of the valve and ensure that it will close in the event of a line failure.

Consequences Avoided: An excess flow valve can prevent a large release caused by a hose or line failure.

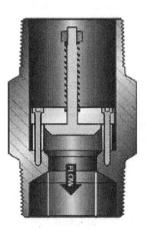

Figure 5.34. Excess flow valve, courtesy of Total Valve Systems.

Data Table 5.29. Excess flow valve

IPL description
Excess flow valve
Generic PFD suggested for use in LOPA
0.1

A thorough analysis may justify a PFD of 0.01 for some systems in clean, nonfouling service. |
| **Special considerations for use of generic PFD for this IPL** |
| • The excess flow valve is properly designed for closure and properly installed in a storage tank outlet or line to ensure that flow from an accidental piping failure downstream of the valve will be stopped.
• Minimizing backpressure downstream of the excess flow valve is critical to ensure that the triggering flow is reached.
• The method of Freeman and Shaw (1988) may be used to verify that a material in a given application could be released at a flow rate that would close an excess flow valve.
• For an incompressible fluid, standard fluid flow calculations, such as those presented in *Flow of Fluids: Through Valves, Fittings and Pipe*, Crane Technical Paper 410 (Crane 2009), can be used to verify that closure of the valve is possible.
The excess flow valve is
• Capable of preventing the scenario of concern
• In clean, noncorrosive, nonfouling service
• Properly installed in a nonvibrating, low-stress area |
| **Generic validation method** |
| The inspection and testing frequency is set according to the manufacturer's recommendation and previous inspection history. ASTM Standard F1802-04 (2010) is a reference that covers the requirements and test methods for excess flow valves in natural gas piping systems. Failure mechanisms of concern include blockage of the valve, corrosion of the valve seats, mechanical damage to the valve internals, and vibration-induced leakage or valve failure. |
| **Basis for PFD and generic validation method** |
| Consensus of the *Guidelines* subcommittee. The principle of operation and design of an excess flow valve is similar to that of a check valve. Therefore, the reliability characteristics are expected to be the same as long as the excess flow valve is managed with a comparable ITPM program (including periodic inspection and testing). |

Restrictive Flow Orifice (RFO)

Description: A restrictive flow orifice (sometimes termed a *restricting orifice*) is used to limit the risk of an excess flow from a source. It is a mechanical restriction of flow diameter with no moving parts. The diameter of a restrictive flow orifice is sized based on the maximum flow rate desired and the expected upstream pressure; if the upstream pressure increases, this maximum flow assumption may no longer be valid. The generic PFD provided in the data table assumes the use of a restrictive flow orifice in clean service for the prevention of an excess flow rate that could result in an undesired consequence.

Consequences Avoided: A restrictive flow orifice prevents flow higher than a specified rate, given a maximum potential pressure drop across the orifice.

Figure 5.35. Diagram of a restrictive flow orifice, courtesy of Energy Project Solutions Ltd.

Data Table 5.30. Restrictive flow orifice

IPL description
Restrictive flow orifice
Generic PFD suggested for use in LOPA
0.01
Special considerations for use of generic PFD for this IPL
• The orifice is in clean, noncorrosive, and nonerosive service and can effectively restrict flow to less than the specified flow rate under all conditions and upstream pressures. Calculations show that the orifice is sized correctly for the specified conditions. • The organization's ITPM and MOC systems prevent removing and modifying (drilling out, enlarging) a restrictive orifice.
Generic validation method
Periodic inspection is performed at a frequency based on the results of previous inspections. Visual examination is performed to detect corrosion, erosion, or fouling of the orifice. Measurement of the orifice opening diameter confirms that it is of the appropriate size.
Basis for PFD and generic validation method
Consensus of the *Guidelines* subcommittee.

Pipeline Surge Dampening Vessel

Description: The dampening vessel is an inline vessel that acts as a shock absorber for flow variations. The PFD provided in Data Table 5.31 assumes that the piping system is designed to avoid pressure surge such that the IPL is being used in low demand mode. If pressure surges are more frequent, then pipeline surge dampening vessels may actually be operating in high demand mode, and the appropriate IEF should be determined. Refer to Section 3.5.2 for information on high demand mode operation.

Consequences Avoided: A pipeline surge dampening vessel prevents the potential consequences of pressure shocks (water hammer) as a result of abrupt flow changes.

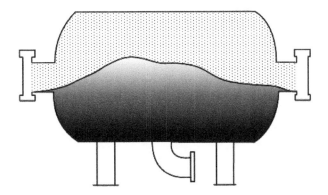

Figure 5.36. Bladder type of surge dampening vessel.

Data Table 5.31. Pipeline surge dampening vessel

IPL description
Pipeline surge dampening vessel
Generic PFD suggested for use in LOPA
0.01 for failure to prevent shock due to pressure shocks (water hammer)
Special considerations for use of generic PFD for this IPL
• The surge vessel is confirmed to be sized for the scenario being considered. • Procedures are in place to maintain the bladder blanket gas at the appropriate pressure.
Generic validation method
Inspection intervals are based on the manufacturer's recommendations and the results of previous inspections. • Inspections are performed to detect the onset of bladder failure. • Replacement of bladders is performed according to the manufacturer's recommendations.
Basis for PFD and generic validation method
Consensus of the *Guidelines* subcommittee.

Check Valves as IPLs

Description: A check valve (also known as a nonreturn valve, one-way valve, or clack valve) is a mechanical device that normally allows liquid or gas to flow through it in only one direction. Check valves work automatically and do not have valve handles or stems. They open in the normal direction of flow at a specific pressure (sometimes called the cracking pressure). Check valves are more effective where there is higher differential pressure across the valve and tend to be less effective in low differential pressure applications. There are various types of check valves used in a variety of applications:

- A *clapper valve* is a type of check valve used in many process industry settings, which has a hinged gate (often with a spring holding it shut until the fluid reaches the cracking pressure).

- A *swing check valve* is a butterfly-style check valve in which a disk swings on a hinge, either onto the seat to block reverse flow or off the seat to allow forward flow. The seat opening cross-section may be perpendicular to the centerline between the two ports or at an angle. Large check valves are often swing check valves.

- A *ball check valve* uses a ball as the moveable part to block the flow. In some ball check valves, the ball is spring-loaded to keep it shut. Ball check valves use conically-tapered surfaces to guide the ball into the seat and form a seal when stopping reverse flow. Similarly, designed check valves use poppets, instead of balls. A *lift-check valve* acts similarly to a ball check valve, except that a disk, sometimes called a *lift*, unseats at higher upstream pressure to allow flow to the downstream side. A guide keeps motion of the disk in-line so the valve can later reseat properly. Lift-check valves are often used in liquid natural gas service.

- A *diaphragm check valve* uses a flexing rubber diaphragm positioned to create a normally closed valve. The pressure on the upstream side must be greater than that on the downstream side for the check valve to allow flow. Once the positive differential pressure stops, the diaphragm automatically flexes back to its original closed position.

- A *tight seal check valve* allows tight shut off against reverse flow. Normally, spring loaded, soft seats (polymeric materials) are used to provide tight shut off.

In *CCPS LOPA* (2001), check valves were not considered to be valid IPLs due to a lack of data supporting their reliability. Since that time, understanding of check valve reliability has improved, assisted by more data that substantiate their reliability. (Refer to Appendix D for Example Reliability Data Conversion for Check Valves for more information.) Therefore, check valves can be considered to be IPLs when properly specified, designed, installed, and maintained.

Check valves are designed to operate with varying degrees of backflow, or leakage. There are a number of standards for leakage rates, such as DIN EN 917

(DIN 1997), which covers thermoplastics valves, BS 6364 (1984), which covers cryogenic valves, and the three standards that are used most in the chemical, oil and gas, and petrochemical industry: API 598 (2009b), ANSI/FCI 70-2 (2006), and MSS-SP-61 (2009). As per ANSI/FCI 70-2 (2006), check valves can be categorized into classes, typically Class I through Class VI, with reduced seat leakage specifications as the class level increases:

Class I – Known as dust tight but with no actual shop test to confirm performance capability. Identical to Class II, III, and IV in construction and design intent, but no actual shop test was made.

Class II – Intended for double port or balanced single port valves with a metal piston ring seal and metal to metal seats. The test medium is air at 45–60 psig.

- Maximum 0.5% leakage of full-open valve capacity
- Tested at service differential pressure or 50 psid (3.4 bar differential), whichever is lower, at 50–125 °F

Class III – Intended for the same types of valves as in Class II. The test medium is air at 45–60 psig.

- Maximum 0.1% leakage of full-open valve capacity
- Tested at service differential pressure or 50 psid (3.4 bar differential), whichever is lower, at 50–125 °F

Class IV – Intended for single port and balanced single port valves with extra tight piston seals and metal-to-metal seats. The test medium is air at 45–60 psig.

- Maximum 0.01% leakage of full-open valve capacity
- Tested at service differential pressure or 50 psid (3.4 bar differential), whichever is lower, at 50–125 °F

Class V – Intended for the same types of valves as Class IV. The maximum leakage limit is 5–10 ml/min per inch of orifice diameter per psi differential.

- Tested at service differential pressure at 50–125 °F
- The test fluid is water at 100 psig or operating pressure

Class VI – Known as a soft seat classification. Soft seat valves are those where the seat or shut-off disc (or both) are made from a resilient material such as Teflon®. The maximum leakage limit depends on valve size and ranges from 0.15 to 6.75 ml/min for valve sizes from 1 to 8 inches.

- The test pressure is the lesser of 50 psig or operating pressure
- The test fluid is air or nitrogen

When specifying a check valve, it is necessary to consider the amount of backflow leakage that can be tolerated before the check valve can no longer be considered an IPL for the scenario of concern. In some scenarios, where a gross backflow is required to cause the consequence, a greater amount of backflow leakage from the check valve can be tolerated. For other scenarios, where a smaller amount of leakage could still allow the incident to occur, a higher class of check valve is specified to allow only minimal seepage. The check valve design should prevent the consequence of concern to claim a check valve as an IPL.

Some companies address the potential for excess check valve leakage by using a second check valve. Using two check valves in series can present challenges, however. Check valves are designed to operate based on differential pressure, and there needs to be sufficient differential pressure in the system to properly operate both check valves. Unless the system design allows for independent functional testing of each check valve in situ, to validate proper operation, removal of the two check valves for bench testing might be necessary. When redundant equipment is used, consideration should also be given to potential common cause failure associated with installation and maintenance.

Care should also be taken when a check valve is used in combination with another device and both devices must function for the system to be effective. One such application is a check valve and pressure relief valve protecting a low pressure system from downstream high pressure. If the relief valve is designed for only a fraction of the potential reverse flow, assuming that the check valve is at least partially effective, the check valve and pressure relief valve are not independent IPLs. Refer to Section 5.2.2.2.2 for more discussion of this topic.

Consequences Avoided: Backflow of material, causing a consequence of concern.

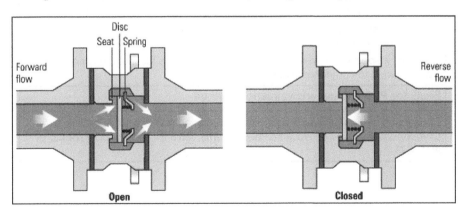

Figure 5.37. Disc check valve (illustrations and text are copyright, remain the intellectual property of Spirax Sarco, and have been used with their kind permission).

Data Table 5.32. Check valve

IPL description
Check valve
Generic PFD suggested for use in LOPA
0.1
Special considerations for use of generic PFD for this IPL
• The check valve operates in low-demand mode; otherwise, refer to Section 3.5.2 regarding high-demand mode operation.   • The service is assumed to be clean, with no fouling or corrosion expected.   • Even properly inspected and maintained check valves might not completely eliminate check-valve seat leakage. Consequently, the user should be aware that isolation of the low-pressure system upstream of the check valve can still result in overpressure.   • It is necessary for the user to define the tolerable leakage rate for the scenario and determine whether a specific check valve is a valid IPL for the scenario of concern.
Generic validation method
The testing or replacement interval is set based on the results of the manufacturer's recommendations and the history of previous inspections.   • Testing in situ or off-line can be conducted by providing back pressure to monitor for leakage.   • Internal inspections can detect onset of failure due to fouling, plugging, corrosion, and component wear.
Basis for PFD and generic validation method
Consensus of the *Guidelines* subcommittee, based on *Reliability Data Book for Components in Swedish Nuclear Power Plants,* RKS/SKI 85-25, p 79 (Bento et al. 1987).

Pressure Reducing Regulator

Description: A pressure regulator is a valve that regulates the flow of a liquid or gas at a certain pressure. These devices reduce the pressure from an upstream source to allow the use of the fluid in a lower pressure downstream system.

Pressure regulators can be used as control devices to maintain a constant pressure in a line. In this application, the pressure regulator is not an IPL; it is acting as a control device. When a pressure regulator is used in a backup capacity, to prevent high pressure if the primary control device fails, then it can be considered as an IPL. Some applications use two pressure reducers in series, with the first reducer being used as the control element and the second reducer being used as an IPL. Often, a pressure gauge is located between the two regulators to detect failure of the primary control element.

Pressure regulators normally have a handle that is adjusted manually to set the downstream pressure at the desired setpoint. Once set, the regulator is normally not readjusted.

Maintenance practices vary widely and can significantly affect the reliability of pressure regulators. With good ITPM practices, an improved PFD may be achieved.

Consequences Avoided: The consequence avoided is a pressure-related process deviation.

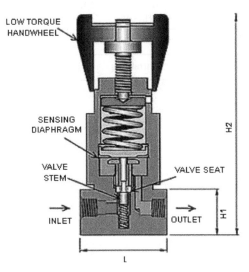

Figure 5.38. Pressure reducing regulator, courtesy of Cunico/Dynamic Controls.

Data Table 5.33. Pressure reducing regulator

IPL description
Pressure reducing regulator
Generic PFD suggested for use in LOPA
0.1, for failure to maintain pressure
Special considerations for use of generic PFD for this IPL
• The pressure reducing regulator operates in low-demand mode; otherwise, refer to Section 3.5.2 regarding high-demand mode operation. • The service is assumed to be clean, with no fouling or corrosion expected. • The spring tension device is manually adjusted to the proper setpoint.
Generic validation method
The inspection interval is set based on the manufacturer's recommendation, the results of previous inspections, and the experience with other regulators in similar service. • In certain services, industry standards may specify maintenance frequencies. • Inspection is done via an independent measurement of controlled pressure. • Visual inspection is performed to detect incipient failure due to corrosion, fouling, or wear on springs, bladders, etc.
Basis for PFD and generic validation method
Consensus of the *Guidelines* subcommittee.

Continuous Pilot

Description: The continuous pilot serves as an independent ignition source for the main burner. A continuous pilot prevents uncombusted material from accumulating in fired equipment if a main burner flame-out occurs. It is important to ensure that the cause of failure of the main burner would not also impact the operation of the continuous pilot. (For example, if the main burner and the continuous pilot share a common fuel source, a continuous pilot would not be an effective IPL against scenarios involving a loss of fuel supply or poor fuel quality.)

A continuous pilot is considered to be a passive IPL. It is a continuously operating device that does not need to respond to a trigger for protection to exist.

Consequences Avoided: A continuous pilot can prevent an explosion within the equipment as a result of the accumulation of flammable material that subsequently ignites.

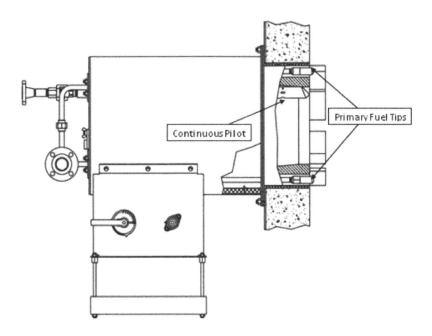

Figure 5.39. Burner with a continuous pilot, courtesy of Zeeco USA, LLC.

Data Table 5.34. Continuous pilot

IPL description
Continuous pilot
Generic PFD suggested for use in LOPA
0.1
Special considerations for use of generic PFD for this IPL
Continuous pilots are provided with an independent, good quality fuel source.
Generic validation method
• Continuous performance of the pilot is validated by monitoring the pilot flame by an independent means. • Preventive maintenance may also be appropriate, if instrumentation is used to monitor the pilot. • Repair of the pilot is initiated when the loss of the pilot flame is revealed.
Basis for PFD and generic validation method
Consensus of the *Guidelines* subcommittee.

Captive Key/Lock System

Description: A captive key/lock system employs mechanical linkages that are released by unique keys to prevent movement of a device (such as door handle or a valve). This prevents humans from operating the valves in the wrong sequence. The captive key lock capability is an integral part of the hardware design and is not able to be removed or defeated by tools readily available to the worker.

Consequences Avoided: A captive key/lock system prevents an initiating event such as opening a valve in the wrong sequence.

> NOTE: A captive key/lock system can also be used to improve the PFD of an IPL, such as a PRV, by ensuring that the IPL is available. In such cases, the captive key/lock system is not a separate IPL; rather, it is included in the PFD of the primary IPL.

Figure 5.40. Captive key device on a ball valve, courtesy of
Castell Interlocks Inc.

Data Table 5.35. Captive key/lock system

IPL description
Captive key/lock system
Generic PFD suggested for use in LOPA
0.01
Special considerations for use of generic PFD for this IPL
This passive IPL prevents incorrect operation of valves or doors that otherwise could lead to an incident. For the captive key/lock system to be effective: • The locking feature is mechanically robust. • The key is released from another mechanical device, and there are no copies of the key readily available to the workers. • Incorrect positioning or improper operation of the captive key/lock system is easily detectable by a visual check.
Generic validation method
• Periodic visual inspection is done to ensure that the device is in the correct position and that safeguards have not been violated. • Audits can be used to confirm that access to any spare keys is controlled.
Basis for PFD and generic validation method
Consensus of the *Guidelines* subcommittee.

Multiple Mechanical Pump Seal System with Seal Failure Detection and Response

A multiple seal system on a pump is a containment system consisting of a primary and secondary seal mechanism with a barrier seal fluid and a means to detect and respond to a primary or secondary seal failure. "Risk Based Pump-Seal Selection Guideline Complementing ISO21049/API 682" (Goodrich 2010) presents an overview of the standard API pump seal systems and a selection methodology for deciding which system to use.

Since almost all seals use the process liquid or gas to lubricate the seal faces, they are, in effect, designed to leak. Process liquids and gases containing hazardous, toxic, or flammable materials are generally not permitted to leak into the atmosphere or onto the ground. In these applications, a secondary "containment" seal is placed after the primary seal along the pump shaft. The space in between these two seals is filled with a neutral or compatible liquid or gas called a "buffer" (unpressurized) or "barrier" (pressurized) fluid.

In a tandem seal (face-to-back), the primary seal will leak into the buffer fluid contained in the unpressurized cavity commonly known as thermosiphon pot. If the cavity registers a significant change in pressure or fluid level, the operator can be made aware that the primary seal has failed. Pressure/level switches or transmitters are commonly used to detect this failure. This arrangement is commonly used with sealing fluids that would create a hazard or change state when in contact with air. This is detailed further in API 682 (API 2002). Once the leak is detected, the process can then be shut down and maintenance can be performed before the secondary seal fails.

In a double seal (generally back-to-back), the barrier liquid in the cavity between the two seals is pressurized. If the primary seal fails, the neutral liquid will leak into the pump stream instead of into the atmosphere. This application is usually used in gas service or with unstable, highly toxic, abrasive, corrosive, and viscous fluids. API 682 (API 2002) provides additional guidance. Typically, nitrogen is used, as its inert nature makes it generally compatible when mixed with the process stream being sealed.

It is possible that factors affecting the primary seal can also affect a secondary seal. Refer to Section 4.3.4.1 for considerations related to factors that can influence pump seal performance.

To be effective, a multiple mechanical seal system requires a means for leak detection when the primary or secondary seal fails, such as a loss of barrier fluid level or pressure. Detection can be performed by means of instrumentation or by visual inspection of an indicator such as barrier fluid level. Response to a primary seal leak can also occur by means of SCAI and/or human intervention. When determining the overall PFD of a double mechanical seal system with leak detection and alarm, it is important to consider the PFDs associated with the detection and the response to the leak. Refer to Data Tables 5.12–5.14 for considerations regarding SCAI and Data Table 5.46 for considerations regarding human response to alarm.

A primary mechanical seal typically has an IEF of 0.1/yr for catastrophic failure; refer to Data Table 4.14. A multiple mechanical seal with adequate leak detection and response acts as an IPL to protect against loss of material from the secondary seal to the environment once the primary seal has failed. For a multiple mechanical seal system with leak detection and response, the IEF of a catastrophic primary seal failure would be 0.1/yr and the IPL PFD of the secondary seal with leak detection and response would be 0.1 PFD, for a combined frequency of 0.01/yr for a failure of the system.

Multiple Mechanical Seal System with Seal Failure Detection and Response

Description: A multiple seal system on a pump is a containment system consisting of a primary and secondary seal mechanism, with a barrier seal fluid and a means to detect and respond to a primary or secondary seal failure.

Consequences Avoided: A multiple mechanical seal system with seal failure detection and response can prevent a leak of material from the pump seal, avoiding the consequences associated with loss of containment.

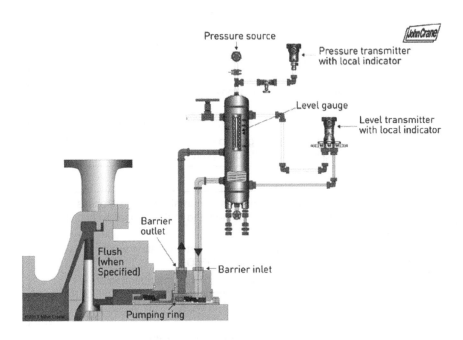

Figure 5.41. Double mechanical seal with seal failure detection and response, © John Crane Group Corporation. All rights reserved. Used with permission.

Data Table 5.36. Multiple mechanical pump seal system with seal failure detection and response

IPL description
Multiple mechanical pump seal system with seal failure detection and response
Generic PFD suggested for use in LOPA
0.1
Special considerations for use of generic PFD for this IPL
<ul><li>The multiple mechanical pump seal system includes at least two mechanical seals, with means to detect and indicate a failure of the primary or secondary seal to the operator. Detection of leakage of one of the seals in a multiple mechanical seal system can be performed by online monitoring of the pressure of the barrier fluid or the level in the tank supplying barrier fluid to the seal. Once a seal leak is identified, the pump is isolated and repaired in a timely manner before another seal fails.</li><li>The operator response to shut down the pump and isolate the source of the leak is documented in a procedure, and the operator is trained to perform the task.</li><li>The task can be completed within the allowable response time. See Section 3.3.3.</li><li>Properly sizing the pump for the application will increase the life of the seal.</li><li>Pump foundation design, piping design and installation, and pump alignment can all significantly impact the seal life on a pump.</li></ul>
Generic validation method
<ul><li>Visual surveillance (such as routine rounds), or an alarm in the process control room, can ensure that seal failure is detected and that action taken prior to a loss of containment.</li><li>Inspections and proof tests of leak detection instrumentation are conducted at an interval necessary to achieve the claimed PFD.</li><li>Inspection and proof testing of the leak detection instrumentation are performed according to a procedure, by personnel who are capable of identifying the as-found/as-left conditions and verifying the integrity and reliability of the equipment.</li><li>For instrumented indication, alarms are documented within the alarm history system.</li><li>Verification of the procedures, training, and control of human factors ensures continuing effectiveness of a human response.</li></ul>
Basis for PFD and generic validation method
Consensus of the *Guidelines* subcommittee.

Continuous Ventilation without Performance Monitoring

Description: Continuous ventilation entails the transfer of a potentially hazardous atmosphere from inside a building or structure to outdoors using fans or blowers. This system is not equipped with performance monitoring equipment, such as alarms to detect low air flow or loss of power to the blower fan.

Consequences Avoided: Potential consequences that may be prevented include asphyxiation, buildup of flammable vapors within an enclosure, and excessive exposure to airborne toxic materials.

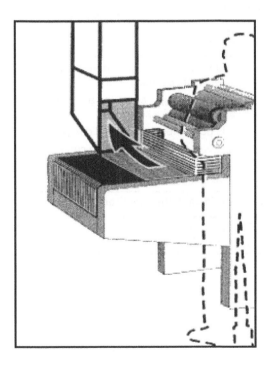

Figure 5.42. Controlling cleaning-solvent vapors at small printers
(CDC 1998).

Data Table 5.37. Continuous ventilation *without* automated performance monitoring

IPL description
Continuous ventilation *without* automated performance monitoring
Generic PFD suggested for use in LOPA
0.1
Special considerations for use of generic PFD for this IPL
This IPL is valid when release into an enclosure or room creates a consequence of concern, and the ventilation is determined to be capable of mitigating the potential hazard. Some useful standards for design and integrity management include the ACGIH *Industrial Ventilation: A Manual of Recommended Practice*, 25th Edition (ACGIH 2004), ANSI-Z9.2 *Fundamentals Governing the Design and Operation of Local Exhaust Systems* (ANSI/AIHA/ASSE 2012), and NFPA 69 *Standard on Explosion Prevention Systems* (NFPA 2008c).
Generic validation method
Periodic inspection and functional testing is performed as per the manufacturer's recommendation and previous inspection history. • Visual inspection shows that fan(s) are running and that dampers are functioning properly and are not blocked. • Periodic maintenance is performed on fans and dampers. • Measurement of air flow velocities versus design can be performed to verify the capacity and distribution throughout the room or enclosure. • Dampers are adjusted as necessary to balance flow.
Basis for PFD and generic validation method
Consensus of the *Guidelines* subcommittee. See *Guidelines for Safe and Reliable Instrumented Protective Systems* (CCPS 2007b), p288, Table B.4.

Continuous Ventilation with Automated Performance Monitoring

Description: Continuous ventilation entails the transfer of a potentially hazardous atmosphere from inside a building or structure to outdoors using fans or blowers.

Some continuous ventilation systems are equipped with performance monitoring systems, such as alarms or interlocks, that detect when the air flow decreases below a preset amount or if the blower fan turns off. This can be effective where a hazardous environment does not immediately develop upon loss of the fan and there is time to repair the fan or shut down the process. The monitoring is intended to improve the reliability of the system by an order of magnitude.

Consequences Avoided: Potential consequences that may be prevented include asphyxiation, buildup of flammable vapors within an enclosure, and excessive exposure to airborne toxic materials.

Figure 5.43. Ventilated lab hood with automated performance monitoring, courtesy of Air Clean Systems.

Data Table 5.38. Continuous ventilation *with* automated performance monitoring

IPL description
Continuous ventilation *with* automated performance monitoring

Generic PFD suggested for use in LOPA
0.01

Special considerations for use of generic PFD for this IPL
This IPL is valid when release into an enclosure or room creates a consequence of concern, and the ventilation is determined to be capable of mitigating the potential hazard. Some useful standards for design and integrity management include the ACGIH *Industrial Ventilation: A Manual of Recommended Practice,* 25th Edition (ACGIH 2004), ANSI-Z9.2 *Fundamentals Governing the Design and Operation of Local Exhaust Systems* (ANSI/AIHA/ASSE 2012), and NFPA 69 *Standard on Explosion Prevention Systems* (NFPA 2008c). Performance monitoring: • The preferred alarm is based on a reliable measurement that ensures that airflow exceeds a minimum value. The alarm functions independently of loss of power to the ventilation fan. • Some performance measurement devices are also interlocked to shut down the operation upon loss of the fan.

Generic validation method
Periodic inspection and functional testing is performed as per the manufacturer's recommendation and previous inspection history. • Visual inspection shows that fan(s) are running and that dampers are functioning properly and are not blocked. • Periodic maintenance is performed on fans and dampers. • Measurement of air flow velocities versus design can be performed to verify the proper capacity and distribution throughout the room or enclosure. • Dampers are adjusted as necessary to balance flow. • Functional test is performed on the instrumentation detecting loss of ventilation.

Basis for PFD and generic validation method
Consensus of the *Guidelines* subcommittee. See *Guidelines for Safe and Reliable Instrumented Protective Systems* (CCPS 2007b), p 288, Table B.4.

Emergency Ventilation Initiated by Safety Controls, Alarms, and Interlocks (SCAI)

Description: A potentially hazardous concentration is detected by instrumented means, often by a toxic or flammable gas sensor. Emergency ventilation is then initiated by an automated interlock, and fans or blowers are used to transfer vapors from the area. An emergency ventilation system initiated by SCAI is designed to detect the component of concern at a low enough level to allow sufficient IPL response time (IRT), and to provide sufficient ventilation capacity once the release is detected, so that the concentration is maintained at a safe level.

A critical issue with automated emergency ventilation is that power is required for ventilation system operation. If the initiating event is caused by loss of power, this IPL may not operate. Also, the frequency at which the facility experiences loss of power would be considered when evaluating the effectiveness of this potential IPL. It is important to demonstrate that the overall ventilation system, including the SCAI equipment, the fan or blower, and the power supply for the system achieves the required performance claim.

Consequences Avoided: Potential consequences that may be prevented include asphyxiation, buildup of flammable vapors within an enclosure (leading to fire or vapor cloud explosion), and excessive exposure to airborne toxic materials over a period of time.

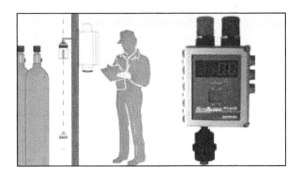

Figure 5.44. Detection of a potentially hazardous atmosphere, courtesy of Sensidyne, LP-www.Sensidyne.com.

Data Table 5.39. Emergency ventilation initiated by safety controls, alarms, and interlocks (SCAI)

IPL description
Emergency ventilation initiated by safety controls, alarms, and interlocks (SCAI)
Generic PFD suggested for use in LOPA
0.1
Special considerations for use of generic PFD for this IPL
This IPL is valid only if the ventilation is determined to be capable of mitigating the potential hazard. Some useful standards for design and integrity management include the ACGIH *Industrial Ventilation: A Manual of Recommended Practice,* 25th Edition (ACGIH 2004), ANSI-Z9.2 *Fundamentals Governing the Design and Operation of Local Exhaust Systems (ANSI/AIHA/ASSE 2012),* and NFPA 69 *Standard on Explosion Prevention Systems* (NFPA 2008). • The overall ventilation system, including the SCAI equipment, the fan or blower, and the power supply for the system achieves the required performance claim. • The instrumentation used to start this emergency system meets the requirements provided in Data Tables 5.13 or 5.14 for either a safety interlock IPL or a SIL 1 loop IPL.
Generic validation method
• Periodic preventive maintenance of fans and dampers is performed based on the manufacturer's recommendations and the results of previous inspections. • Measurements of air flow velocities versus design are done to verify the system capacity and the distribution throughout the room or enclosure. • Dampers are adjusted as necessary to balance the flow. • The instrumented detection system is maintained; refer to Data Tables 5.13 and 5.14 for more information on safety interlocks and SIS.
Basis for PFD and generic validation method
Consensus of the *Guidelines* subcommittee.

Mechanical Interlocks

There are many types of mechanical linkages and interlocks that may qualify as IPLs or improve the PFD of other IPLs. These include a linkage to ensure that at least one relief valve on a vessel is always in service and a linkage to ensure that a spool piece is in place before opening a downstream valve. Many of these are custom designed, so a general specification, IPL description, and related PFD cannot be assigned. In some cases, the PFD for the mechanical feature may have already been included in the PFD of another IPL, such as the linkage that ensures that a relief device is always in service. Analysts wishing to use such devices can derive IPL descriptions and PFD values for use within their companies.

The following are examples of mechanical interlocks and mechanical features that are now in common use and where generic IPL descriptions and PFD values can be provided.

Mechanically Activated Emergency Shutdown/Isolation Device

Description: This type of emergency shutdown device (ESD) has a mechanical linkage that actuates an isolation mechanism (such as a spring-loaded valve or a valve that vents the instrument air of an air-to-open valve). The linkage activates mechanically in response to a deviation. A mechanical stop or block can also be engaged once a mechanical feature is set (such as a locking feature on a key-lock valve or switch).

> *Example for rail car line isolation: Pins hold an isolation valve (or multiple isolation valves) open on the rail car and process sides of an unloading hose connection. If the rail car to which the pins are attached moves too far, the pins are pulled out by a mechanical linkage and the isolation valves close.*

Consequences Avoided: The device isolates a flow path to limit a release.

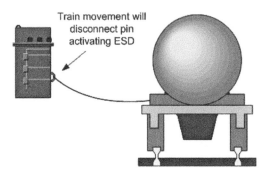

Figure 5.45. Example of a mechanically activated shutdown device.

Data Table 5.40. Mechanically activated emergency shutdown/isolation device

IPL description
Mechanically activated emergency shutdown/isolation device
Generic PFD suggested for use in LOPA
0.1
Special considerations for use of generic PFD for this IPL
The IPL is designed to meet the requirements of the specific system.
Generic validation method
• Periodic inspection is performed to ensure the integrity and operation of the sensing component and final element. • Field auditing can help ensure that the device is being used properly.
Basis for PFD and generic validation method
Consensus of the *Guidelines* subcommittee.

Mechanical Overspeed Trip on a Turbine

Description: The mechanical overspeed trip on a turbine is a device that operates to stop the turbine when its speed exceeds a predetermined value. Typically, a trip bolt is set slightly off-center and is held in place by a spring force; as speed increases and exceeds the over-speed limit, centrifugal force on the bolt exceeds the spring force. The bolt extends, striking the trip paddle and releasing the linkage.

Modern overspeed protection systems often use SCAI rather than a mechanical overspeed trip. In these cases, refer to the appropriate SCAI data tables for the specific configuration.

Consequences Avoided: Mechanical overspeed protection averts catastrophic destruction of rotating equipment.

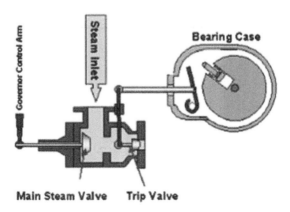

Figure 5.46. Mechanical overspeed trip, courtesy of
Dresser-Rand Company, Olean, NY.

Data Table 5.41. Mechanical overspeed trip on a turbine

IPL description
Mechanical overspeed trip on a turbine
Generic PFD suggested for use in LOPA
0.1
Special considerations for use of generic PFD for this IPL
This IPL represents a fully mechanical machine protection system. It usually consists of a bolt, a tripping mechanism that drains the control oil, and the trip and throttle (or stop) valve. • Other types of overspeed protection devices consist of sensors, a logic solver, and final elements. These are considered to be SCAI and are credited as a safety interlock or a SIS IPL. Refer to Data Tables 5.13 and 5.14.
Generic validation method
• Inspections of components are performed for wear, corrosion, or erosion. • Preventive maintenance is done to ensure reliability. • The turbine speed is raised and lowered, and the system is adjusted until the interlock trips at the appropriate setpoint. Normally, this is done with the driven unit uncoupled to avoid unnecessary stress caused by high rotating speeds.
Basis for PFD and generic validation method
Consensus of the *Guidelines* subcommittee.

Fire and Explosion Suppression Systems

Fire and explosion suppression systems are used to stop events within process equipment and to mitigate events outside of process equipment. Below are typical IPLs for mitigating fires and explosions within process equipment.

The following four types of systems are discussed in this section:

1. Automated fire suppression system (within process equipment)
2. Automated fire suppression system for local application
3. Automated fire suppression system for a room
4. Automated explosion suppression system within process equipment

> NOTE: The suppression system is dependent on proper location of the sensors to detect the fire or smoke condition. If the fire or smoke does not contact the sensor, the suppression system cannot operate effectively.

Automated Fire Suppression System (within Process Equipment)

Description: The fire suppression system works to extinguish a fire within a piece of process equipment and to prevent spread of the fire. Types of automatic fire protection systems include water, water and foam, and other suppressants. These systems typically use fire or smoke detectors to automatically activate a system designed to prevent or control a fire.

Consequences Avoided: The automated fire suppression system prevents propagation of a fire outside of process equipment.

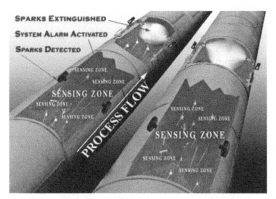

Figure 5.47. Spark detection and extinguishing, courtesy of BS&B Safety Systems, Limited.

Data Table 5.42. Automatic fire suppression system (within process equipment)

IPL description
Automatic fire suppression system (within process equipment)
Generic PFD suggested for use in LOPA
0.1
Special considerations for use of generic PFD for this IPL
A number of standards can provide guidance on automatic fire suppression systems for process equipment: • NFPA 11 *Standard for Low-, Medium-, and High-Expansion Foam* (2010b) provides guidance on foam systems used in storage tanks. • NFPA 69 *Standard on Explosion Prevention Systems* (2008c) provides guidance on the use of suppression systems in process equipment. • NFPA 91 *Standard for Exhaust Systems for Air Conveying of Vapors, Gases, Mists, and Noncombustible Particulate Solids* (2010a) provides guidance on automatic extinguishing systems in process equipment. • NFPA 654 *Standard for the Prevention of Fire and Dust Explosions from the Manufacturing, Processing, and Handling of Combustible Particulate Solids* (2013) provides guidance on solids handling systems. • DIN EN 14373:2006 (2006) is the European standard for explosion suppression systems. • ISO 6184-4 *Explosion protection systems – Part 4: Determination of Efficacy of Explosion Suppression Systems* (1985) is the ISO standard covering these systems.
Generic validation method
• NFPA 25 *Inspection, Testing and Maintenance of Water-Based Fire Protection Systems* (2008b) is a useful reference when determining what components to test and identifying test methods. • NFPA 69 *Standard on Explosion Prevention Systems* (2008c) gives guidance on the installation, inspection, and maintenance of suppression systems. • Additional information for other extinguishing systems can be found in applicable NFPA standards.
Basis for PFD and generic validation method
Consensus of the *Guidelines* subcommittee. Refer to *Guidelines for Process Equipment Reliability Data, with Data Tables* (CCPS 1989), p 207. Also, see *Guidelines for Safe and Reliable Instrumented Protective Systems* (CCPS 2007b), p 298, Table B.5.

Automatic Fire Suppression System for Local Application

Description: Automatic fire protection systems for local applications are designed to direct a fire suppressing agent to a specific area of concern, such as a dike or a specific piece of equipment. Typical flooding systems use dry powder, clean agent, carbon dioxide, and recent replacements for Halon agents. NFPA Standard 2001 (2012) addresses clean agent use, including the design, installation, and maintenance requirements for systems that employ replacement agents for Halon 1301.

Consequences Avoided: Fire suppression systems for local application mitigate fires in small areas.

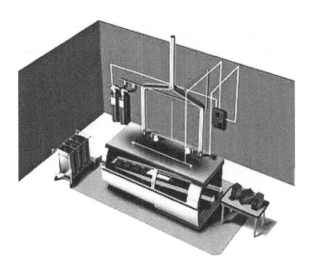

Figure 5.48. Fire suppression system for local application, courtesy of Fike Corporation.

Data Table 5.43. Automatic fire suppression system for local application

IPL description
Automatic fire suppression system for local application
Generic PFD suggested for use in LOPA
0.1
Special considerations for use of generic PFD for this IPL
Refer to • NFPA 17 *Standard for Dry Chemical Extinguishing Systems* (2009) • NFPA 2001 *Standard on Clean Agent Fire Extinguishing Systems* (2012) for information on dry powder clean agent, and other flooding systems
Generic validation method
Refer to • NFPA 17 *Standard for Dry Chemical Extinguishing Systems* (2009) • NFPA 2001 *Standard on Clean Agent Fire Extinguishing Systems* (2012) for guidance on inspection, testing, and maintenance
Basis for PFD and generic validation method
Consensus of the *Guidelines* subcommittee.

Automatic Fire Suppression System for a Room

Description: This type of automatic fire suppression system is designed to extinguish a fire in an enclosure or small room, such as a controlled-environment room housing key computer equipment. Typical flooding systems use dry powder, clean agent, carbon dioxide, and recent replacements for Halon agents. NFPA Standard 2001 (2012) addresses clean agent use, including the design, installation, and maintenance requirements for systems that employ replacement agents for Halon 1301.

Consequences Avoided: Fire suppression systems mitigate fire in a room or small enclosure.

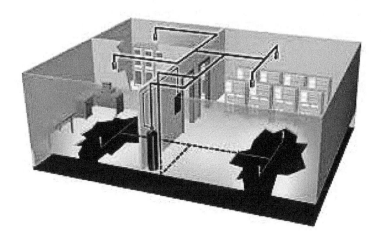

Figure 5.49. Automatic fire suppression system for a room, courtesy of Fike Corporation.

Data Table 5.44. Automatic fire suppression system for a room

IPL description
Automatic fire suppression system for a room
Generic PFD suggested for use in LOPA
0.1
Special considerations for use of generic PFD for this IPL
Refer to • NFPA 17 *Standard for Dry Chemical Extinguishing Systems* (2009) • NFPA 2001 *Standard on Clean Agent Fire Extinguishing Systems* (2012) for information on dry powder, clean agent, and other flooding systems
Generic validation method
Refer to • NFPA 17 *Standard for Dry Chemical Extinguishing Systems* (2009) • NFPA 2001 *Standard on Clean Agent Fire Extinguishing Systems* (2012) for guidance on inspection, testing, and maintenance
Basis for PFD and generic validation method
Consensus of the *Guidelines* subcommittee. See *Guidelines for Process Equipment Reliability Data, with Data Tables* (CCPS 1989), p 209.

Automatic Explosion Suppression System for Process Equipment

Description: Explosion suppression systems work by early detection and rapid action to stop the propagation of an incipient internal deflagration, before the pressure rise becomes potentially damaging to the enclosure being protected. Sensing devices can include thermal radiation, pressure, and/or flame detectors. A signal passes to a control unit that actuates a system to rapidly inject an extinguishing agent. This stops the flame from propagating and prevents further pressure rise within the equipment. Explosion suppression systems generally contain a single charge of dispersant and rely on the subsequent shutdown of the process to prevent a reoccurrence of the scenario.

Automated explosion suppression systems are generally carefully engineered and may include self-checking diagnostics and other design features that improve the system integrity. More quantitative analysis may support a lower PFD value for a specific system than the generic PFD provided in the following data table.

Consequences Avoided: The explosion suppression system protects against explosions that could cause equipment damage, including rupture.

NOTE: Since some deflagration suppression systems use pressure sensors for activation, it may be necessary to disarm the suppression system during cleaning activities of the process equipment to avoid accidental discharge. Thus, it is critical to employ a strong procedure for impairing the safety system and returning it to service to ensure that the suppression system is available during operation.

Figure 5.50. Valve for an automatic explosion suppression system, courtesy of BS&B Safety Systems, Limited.

Data Table 5.45. Automatic explosion suppression system for process equipment

IPL description
Automatic explosion suppression system for process equipment

Generic PFD suggested for use in LOPA
0.1
More quantitative analysis may support a lower PFD value for a specific system than the generic PFD provided.

Special considerations for use of generic PFD for this IPL
The suppression system is mounted to the system to be protected and propels the extinguishing agent. • To be effective, the system is rapidly actuated. • Refer to NFPA 69 *Standard for Explosion Prevention Systems* (2008c) for additional guidance.

Generic validation method
The inspection frequency is set based on the manufacturer's recommendations and past inspection history. • Inspection and testing can be performed in situ or off-line. This includes an internal inspection for fouling, plugging, sticking, and corrosion.

Basis for PFD and generic validation method
Consensus of the *Guidelines* subcommittee.

Fusible Link Devices

Description: A fusible link is a device that is designed to melt at a specific temperature, causing a link to open and break an electric circuit or allow a mechanical linkage to function. Fusible links are used as triggering devices in fire sprinkler systems and mechanical automatic door-release mechanisms that close fire doors in warehouses. One common application is the use of a heat-sensitive fail-closed valve on the discharge of a vessel that restricts the flow of flammable liquids or gases in the event of a fire. Fusible link applications can be found at facilities such as tank farms, airports, and chemical processing plants.

In addition, fusible link assemblies are used as fail-open valves in fire safety protection systems. Fire suppression systems can incorporate a fusible link assembly where the automatic opening of a valve would allow the flow of water or fire-suppressing chemicals to extinguish a fire or cool a storage tank.

Fail-open valves with fusible links are also used in applications designed to protect gas-filled vessels in the event of an external fire. Use of a fail-open valve in this application allows the vessel to be depressurized prior to vessel wall failure. For further guidance on the technical issues around vessel depressurization, refer to ANSI/API 521 (2008).

Fusible links generally require a significant heat source, such as a fire, to function. Since the fire consequence has already occurred, the fusible link serves to mitigate the potential severity of the outcome. It may be necessary to employ other quantitative risk assessment techniques to evaluate the mitigative impact of a fusible link device. Chapter 6 provides additional guidance regarding the use of more quantitative risk assessment techniques.

Figure 5.51. A type of fusible link used in fire suppression systems, courtesy of Globe Technologies Corporation.

5.2.2.4 Vendor-Installed Safeguards

Many equipment items are supplied with various safeguards designed by the equipment vendor. Examples include

- *Fired Equipment* – Burner management systems (BMS) are instrumented systems. They include permissives and interlocks that prevent burner misoperation and reduce risk of uncombusted material being introduced into the fired equipment. BMS are usually separate from combustion controls. Combustion controls serve to adjust burner rates, fuel air ratios, etc. In contrast, BMS are typically made up of discrete loops that provide safeguards for burner operation. For example, if a flame sensor does not detect a flame, a SIS may close a fuel valve. For a scenario involving a potential explosion in a boiler due to fuel gas accumulation, the BMS could be an IPL if properly designed, installed, and maintained.

- *Rotating Equipment* – Vibration switches, high-temperature detection, rotating speed detection and overspeed protection, and anti-surge protection are safeguards that can be found on types of rotating equipment. For some scenarios involving rotating equipment, vendor-supplied interlocks could be IPLs if properly designed, installed, and maintained.

It is appropriate to consider such devices as IPLs for the purposes of LOPA as long as the core attributes of an IPL are met. It is the end user's responsibility to ensure that vendor-supplied systems meet the criteria of an IPL. Factors that would influence this decision and the PFD value include:

- The design of the IPL. (The interlock provided on vendor-supplied equipment meets the requirements for SCAI, with a risk reduction sufficient to support the IPL credit being claimed.)
- Historical data. (Data may be available from the vendors and are reviewed to ensure that the safeguard is sufficiently reliable.)
- The IPL is operated and maintained in a manner that ensures its ongoing reliability at the expected performance level.

5.2.2.5 Human Action IPLs

Human IPLs rely on operators or other staff to take action to prevent an undesired consequence, either in response to an alarm, following a routine check of the system, or while performing verification checks that are part of an established operating procedure. The effectiveness of humans in performing routine and emergency tasks has been the subject of several publications, most notably *Handbook of Human Reliability Analysis with Emphasis on Nuclear Power Plant Applications, Final Report,* NUREG CR-1278 (Swain and Guttmann 1983) and *Human Reliability and Safety Analysis Data Handbook* (Gertman and Blackman 1994).

Chapter 3 defines the IPL Response Time (IRT) as the time necessary for the IPL to respond to the setpoint and take the necessary action. For a human response IPL, the IRT includes the time it takes a sensor to detect the process deviation plus the time for the individual to be alerted to the deviation, diagnose the problem, and correctly execute the required action to bring the process to a safe state.

Operators often take control actions through the BPCS. Normal process control actions are generally not considered to be safety functions. When the operator is expected to take action in response to an alarm to prevent a process safety incident, the alarm and associated operator action are considered to be a safety function. These alarms are designed and managed in a manner that supports the overall PFD of the response. For more information, refer to Section 5.2.2.1.1 regarding requirements for safety alarms.

Although some abnormal conditions are annunciated using alarms, others are not. Conditions that trigger a human response IPL can include field readings or sample results. Likewise, not all human responses are enacted using the BPCS or other instrumented system. A human IPL can involve manually closing a valve following an alarm. A human IPL can also consist of the correct performance in completing a prescribed sequence of steps to return the process to a safe state.

The general requirements for crediting human actions as an IPL are the same as other IPLs discussed in this chapter but are often described in different terms. Human IPLs typically have the following characteristics:

- The indication of a process deviation and the response action are independent of any alarm, instrument, SIS, or other system already credited as part of the initiating event sequence or another IPL.
- The indication for action required by the operator is detectable and unambiguous. The indication is
  - Available to the operator
  - Clear to the operator even under emergency conditions
- There is sufficient time available to successfully complete the required response. Generally, the longer the amount of time available for action, the better the chances of success of a human IPL.
- The decision-making by the operator requires no complicated calculations or diagnostics.
- There is a procedure or troubleshooting guideline available that includes
  - A description of the hazard scenario being prevented
  - The "never exceed, never deviate" process conditions that require manual shutdown or other action to bring the system to a safe state
  - The steps needed to correct or avoid the process deviation. These may be listed as prescriptive steps for a straightforward action, or information may be provided in the form of a troubleshooting guide.

- If the alert is from a SCAI, information available includes
  - A description of the operation of the SCAI
  - The alarm setpoint or interlock setpoint and the expected safe state when a trip is completed
  - Conditions under which it is safe to reset the system
- Verification by the operator that the safe state is achieved
- The normal workload allows the operator to be available to complete the necessary action as part of the human response IPL. During the response, the operator is able to address the abnormal situation in the absence of other demands that might otherwise make the response ineffective.
- The operator is capable of taking the action required under all conditions that could reasonably be expected to be present. If the initiating event would impede the individual from performing the required action, it would not be appropriate to consider human response as an IPL.
- Operators are trained on the procedure that they are expected to follow to respond to the abnormal situation.
- The operator taking the corrective action can do so without being put into a dangerous situation to accomplish the action.
- Human factors related to communications, human-system interface, and the work environment are appropriately managed; refer to Appendix A for a more detailed discussion of this topic.

PFD Determination for Human IPLs

Information is often available on the failure rates of equipment, instruments, and other equipment for which PFD values for component-based IPLs can be established. However, developing PFD values for human IPLs is not as straightforward. A human IPL PFD value includes not only the probability of failure of an incorrect or ineffective human response to an alarm or other trigger, but also the PFDs of instrumentation or equipment used to detect the process deviation and/or take action to return the process to a safe state. To claim a human IPL PFD of less than 0.1, in addition to the factors discussed above, equipment and instrumentation meet the requirements of ANSI/ISA 84.00.01 (2004) or IEC 61511 (2003) for SIL 2 integrity. A human reliability assessment (HRA) is often useful to help determine whether a PFD of 0.01 can be claimed for the combination of the human IPL and the instrumented portion.

The following data tables are example human IPLs and their related PFDs:

Data Table 5.46. Human response to an abnormal condition

Data Table 5.47. Human response to an abnormal condition with multiple indicators and/or indicators, and the operator has > 24 hours to accomplish the response action

Human Response to an Abnormal Condition

Description: The individual responds to an unambiguous cue, such as a safety alarm annunciation, clearly displayed indication of an abnormal condition, or field sample or reading that is outside of predefined safe operating limits. Any hardware/software associated with the detection, annunciation, and response is also sufficiently reliable to achieve the PFD claimed.

There is a procedure or guideline available that indicates the action(s) to be taken, and the operator is trained on the procedure. The steps in the procedure are of low complexity, with clear, step-by-step instructions that are well within the capability of the worker to perform and do not require abstract decision-making. The abnormal condition is detected early enough such that the operator has sufficient time to take the steps necessary to return the process to a safe state. Human factors associated with this response are sufficiently optimized; refer to Appendix A for additional information.

For the IPL to be effective, the operator needs to have sufficient time to respond to the abnormal condition. To credit the human response as an IPL for LOPA, the decision and response time allotted to the operator is long enough to accomplish the task but must be less than the time that it takes for the event to become unavoidable. See Section 3.3.3.

Consequences Avoided: Human response to an abnormal condition can prevent a variety of possible consequences of concern.

Data Table 5.46. Human response to an abnormal condition

IPL description
Human response to an abnormal condition
Generic PFD suggested for use in LOPA
0.1
Special considerations for use of generic PFD for this IPL
<ul><li>When the trigger for the human response is a safety alarm, the alarm is clearly understandable and available to the operators in their usual work location(s).</li><li>When the trigger for the human response is a check or field sample, a procedure indicates the need for this check or sample and the required frequency. There is also written guidance on what to do if the check shows the reading to be out of a tolerable range. Readings are recorded in a checklist, on an appropriate form, or in some form of database.</li><li>The operator has sufficient time available to respond to the indication of an abnormal condition and complete the required action, and this time is less than the time that it takes for event to become unavoidable. See Section 3.3.3.</li><li>There are clear procedures for the operator to follow to complete the response. The response task is low-complexity, with step-by-step instructions and minimal diagnostics or calculations.</li><li>The operator is trained on the response task.</li><li>The operator taking the corrective action can do so without being put into a dangerous situation to accomplish the action.</li><li>The human factors related to oral communication, human system interface, and work environment have been reasonably optimized. (See Appendix A for performance shaping factors to control.)</li></ul>
Generic validation method
<ul><li>Testing of any sensors, alarms, and final control elements used by the operator in the response procedure is performed to ensure that they function properly. (See the Safety Alarm Section 5.2.2.1.1 for details.)</li><li>Verification of the procedures, training, and control of human factors ensures continuing effectiveness of human response.</li><li>Tabletop exercises, drills, and use of a process simulator are techniques that can be used to provide refresher training or to demonstrate response effectiveness.</li></ul>
Basis for PFD and generic validation method
Consensus of the *Guidelines* subcommittee. See NUREG CR-1278 – *Handbook of Human Reliability Analysis with Emphasis on Nuclear Power Plant Applications, Final Report*, Table 15-3 (Swain and Guttmann 1983).

Human Response to an Abnormal Condition with Multiple Sensors and/or Indications, and the Operator Has > 24 Hours to Accomplish the Required Response Action

Description: For this IPL, the individual responds to deviation indications from multiple sensors and/or indications. There are at least 24 hours of response time available, the operator is trained to take immediate action, and the operator has sufficient time to complete the required action necessary to prevent the consequence of concern.

Any hardware/software associated with the detection, alarm annunciation, and response is also sufficiently reliable to achieve the PFD claimed. To claim a human IPL PFD of less than 0.1, in addition to the factors discussed above, equipment and instrumentation are expected to meet the requirements of ANSI/ISA 84.00.01 (2004) or IEC 61511 (2003) for SIL 2 integrity.

There is a procedure or guideline available that indicates the action(s) to be taken, and the operator is trained on the procedure. The abnormal condition is detected early enough such that the operator has sufficient time to take the steps necessary to return the process to a safe state. The operator taking the corrective action can do so without being put into a dangerous situation to accomplish the action. Human factors associated with this response are sufficiently optimized; refer to Appendix A for additional information.

Consequences Avoided: Any scenario of concern where human response can be an effective IPL.

Data Table 5.47. Human response to an abnormal condition with multiple indicators and/or sensors, and the operator has > 24 hours to accomplish the required response action

IPL description
Human response to an abnormal condition with multiple indicators and/or sensors, and the operator has > 24 hours to accomplish the required response action
Generic PFD suggested for use in LOPA
0.01
Special considerations for use of generic PFD for this IPL
To achieve the generic PFD for this IPL: • There are multiple, unambiguous cues that there is an abnormal condition. • The available time for the operator to respond to the alarm and complete the required action is less than the time that it takes for event to become unavoidable. See Section 3.3.3. • There are clear procedures for the operator to follow to complete the response. • The operator is trained on the response task. • The operator taking the corrective action can do so without being put into a dangerous situation to accomplish the action. • The human factors related to oral communication, human system interface, and work environment have been reasonably controlled. (See Appendix A for performance shaping factors to control.)
Generic validation method
• Testing is conducted of the sensor, alarm, and final control element used by the operator in the response procedure to ensure that they function properly. (See the Safety Alarm Section 5.2.2.1.1 for details.) • Verification of the procedures, training, and control of human factors ensures continuing effectiveness of human response. • Field or tabletop drills can be useful means of providing refresher training.
Basis for PFD and generic validation method
Consensus of the *Guidelines* subcommittee. See NUREG CR-1278 – *Handbook of Human Reliability Analysis with Emphasis on Nuclear Power Plant Applications, Final Report*, Table 15-3 (Swain and Guttmann 1983).

Adjustable Movement-Limiting Device

Description: The adjustable movement limiting device prevents human misoperation of equipment beyond a limit that would initiate a scenario of concern. Unlike mechanical stops, which are secured in place and not moved, it is more likely that adjustable movement-limiting devices can be relocated, removed, or otherwise defeated. Examples of adjustable movement-limiting devices include

- A car seal, which is crimped around the ends of a wire that secures a valve in its correct position. The car seal provides a visual inspection aid to ensure that tampering with the valve has not occurred. Note, car seals need to be able to withstand ambient conditions, including sun, temperature, chemicals in the local atmosphere, etc. Plastic ties may not have sufficient integrity to perform the function in some environments.

- A chain and lock system, which requires a single key to unlock the chain and allows the valve position to be changed. The generic PFD value provided in the data table assumes that the key is under supervision and there is a system in place to control access.

- A mechanical stop, which limits travel of a component (such as a piston or valve) or machine (such as a conveyor or rail car).

Consequences Avoided: An adjustable movement-limiting device prevents consequences that arise from the unintended movement of equipment.

NOTE: This IPL does not apply to misalignment of a device that restricts the operation of block valve under a relief valve. Inherent in the PFD values in the relief device data tables is a requirement that there be a management system in place to ensure the proper positioning of these valves, and adjustable movement-limiting devices may already be part of that management system. Refer to Section 5.2.2.2.1 for a discussion of relief valves with associated block valves.

Figure 5.52. Adjustable rail car stop.

Data Table 5.48. Adjustable movement-limiting device

IPL description
An adjustable movement-limiting device, such as a strong wire car seal, chain/lock, or an adjustable mechanical stop, that is intended to prevent operation of a device or movement of an object beyond a defined limit.
Generic PFD suggested for use in LOPA
0.1
Special considerations for use of generic PFD for this IPL
The human or work group that positioned the adjustable movement-limiting device is not involved in the initiating event or other IPL in the same scenario. The movement-limiting device has a means to indicate that only an authorized person can move or actuate the device.
To be effective:
• The device is designed to protect against the consequence of concern.
• The use of adjustable movement-limiting devices is specified in a procedure, and affected individuals are trained on the use of that procedure.
• Individuals are trained to install and secure the movement-limiting device using wires or other seals that have sufficient integrity to perform in the specific operating environment.
• The device is routinely used each time the task is performed.
• There are means to clearly communicate to other personnel that the device is not to be tampered with unless the person is authorized to do so.
• Incorrect positioning of the movement-limiting device is easily detectable by a visual check.
• The equipment is inspected, tested, cleaned, and maintained as needed to ensure its ongoing effectiveness.
• Human factors related to the work environment have been reasonably controlled; refer to Appendix A for more details.
Generic validation method
Periodic visual inspection ensures that the device is in the correct position and the lock/seal/stop has not been violated.
Basis for PFD and generic validation method
Consensus of the *Guidelines* subcommittee.

Personal Protective Equipment (PPE)

Description: Personal protective equipment (PPE) consists of safety gear that is worn to act as a barrier between a person and hazard. Examples of PPE include respirators, chemical suits, face shields, and gloves. Generally, PPE is considered to be a last line of defense. Use of inherently safer principles to eliminate the hazard or the application of engineering controls to reduce the likelihood of the event or magnitude of the material/energy release are preferred to the use of administrative controls. However, there are situations where engineering controls are not feasible, and PPE can be an effective IPL.

Many companies require basic PPE, such as safety shoes, safety glasses, and hard hats, to enter operating areas. Basic PPE is not generally considered to be an IPL since it is not likely to protect against a specific LOPA scenario. To consider PPE as an IPL,

1. The PPE is designed to protect against the specific consequence of concern. If the hazard to be protected against is a significant chlorine leak, for example, use of a cartridge respirator may be inadequate to provide sufficient respiratory protection against the scenario; however, an airline respirator may provide adequate short-term protection. Likewise, goggles would not offer full protection against a liquid exposure to the face but might afford protection against an eye injury, if the goggles are properly designed, fitted, and worn. To be an effective IPL, the PPE is specified to provide full protection against the specific potential hazard of concern.
2. Personnel are trained on the proper use of PPE that they are required to use.
3. The PPE is cleaned, maintained, tested, and replaced as needed. PPE will only be effective if it is functioning as designed.
4. The PPE is worn each time that the task is performed. Tasks that require PPE as part of an IPL are described in a written operating procedure, and the PPE required is also included in the facility's hazard assessment. Affected personnel are trained on the procedure.
5. The PPE is put on before the task is performed. PPE that is donned in response to an event would generally not be considered to be an IPL. It is possible that the individual could be incapacitated or exposed prior to putting the PPE on, and the stress that an event could generate would reduce the effectiveness of a human response.
6. The individual is trained on how to respond if an exposure does occur while wearing the PPE. This may include rinsing off in a safety shower and being medically evaluated. Once the routine PPE is contaminated, it is no longer used to respond to the event. In this case, proper emergency response protocols are followed.

Consequences Avoided: PPE prevents consequences associated with exposure of people within the area of potential impact to a hazard of concern.

Data Table 5.49. Personal protective equipment (PPE)

IPL description
Personal protective equipment (PPE)

Generic PFD suggested for use in LOPA
0.1

Special considerations for use of generic PFD for this IPL
To consider PPE as an IPL: • The PPE is specifically designed for the task and potential hazard, and it will provide full protection against the consequence of concern. • The operator is trained on the proper use of PPE. • The PPE is cleaned, inspected, maintained, and replaced as needed. • The operator wears PPE every time that the procedure is executed. The use of the PPE is required in a procedure, and the specific PPE to be used is listed. Affected personnel are trained on the procedure. • PPE is worn each time the task is performed. • The individual dons the PPE prior to starting the task. • The operator understands what measures to take if PPE is contaminated by chemical exposure.

Generic validation method
• PPE is determined to be suitable protection against the hazard of concern. • Inspection, testing, and maintenance of the PPE ensure its ongoing integrity. • Verification of the procedures, training, and control of human factors ensures that PPE is properly used.

Basis for PFD and generic validation method
Consensus of the *Guidelines* subcommittee.

5.2.2.6 *Response Safeguards*

There are many effective safeguards that lower the risk of an incident by limiting the consequence of the scenario. For example, emergency response safeguards are initiated following a release. Examples include

- Employee warnings (such as plant sirens, horns, or lights) that indicate that an emergency is occurring
- Community notification (such as automatic phone calling or texting systems, neighborhood alarms/horns, and police efforts to notify the public)
- Shelter in place (where neighbors are instructed to remain indoors, and workers take shelter in protected areas)
- Internal emergency response capability (including rescue, firefighting, and hazardous material (HAZMAT) response)
- External emergency responders (such as local fire brigades)

Although these safeguards represent important features and practices, they are not generally used as IPLs in LOPA because their implementation is highly dependent on local conditions, and generic IPL values cannot be developed. Evaluating the effectiveness of emergency response safeguards would likely require the use of more detailed risk assessment techniques. Refer to the more quantitative risk assessment methods discussed in Chapter 6 for additional information.

5.3 WHAT IF YOUR CANDIDATE IPL IS NOT SHOWN IN A DATA TABLE?

The *Guidelines* subcommittee reviewed many candidate IPLs and decided which met the criteria for IPLs. The subcommittee also determined if there were sufficient data available to support generic PFD values for candidate IPLs. As mentioned earlier, many companies, including those that participated on the writing of this guideline, use IPLs that are not listed in Chapter 5.

A company may choose to use IPLs that are not listed in a data table in Chapter 5. If so, it is recommended that

- Data are used from a documented reliable source, or site-specific data is used to support the PFD selected. Refer to Appendices B and C for more guidance on data collection.
- There is compliance with the general requirements for implementing and maintaining an IPL.
- The company has management systems in place to support the initial and ongoing integrity and reliability of the IPL.

It is suggested that companies publish their candidate IPL and supporting data so that others can peer review the data/criteria and potentially benefit from use of the IPL.

REFERENCES

ACGIH (American Council of Governmental Industrial Hygienists). 2004. *Industrial Ventilation: A Manual of Recommended Practice.* 25th Edition. Cincinnati: ACGIH.

ANSI/AIHA/ASSE (National Standards Institute/American Industrial Hygiene Association/American Society of Safety Engineers). 2012. *Fundamentals Governing the Design and Operation of Local Exhaust Ventilation Systems.* Z9.2. Falls Church, VA: AIHA.

ANSI/API (American National Standards Institute/American Petroleum Institute). 2008. *Guide for Pressure-Relieving and Depressuring Systems: Petroleum petrochemical and natural gas industries-Pressure relieving and depressuring systems,* 5th Edition addendum. (ISO 23251 Identical). API Standard 521. Washington, DC: API.

ANSI/FCI (American National Standards Institute/Fluid Controls Institute). 2006. *Control Valve Seat Leakage,* ANSI/FCI 70-2-2006. Cleveland: Fluid Controls Institute, Inc.

ANSI/ISA (American National Standards Institute/International Society of Automation). 2004. *Functional Safety: Safety Instrumented Systems for the Process Industry Sector – Part 1: Framework, Definitions, System, Hardware and Software Requirements.* 84.00.01-2004 (IEC 61511-1 Mod). Research Triangle Park, NC: ISA.

ANSI/ISA. 2009. *Management of Alarm Systems for the Process Industries.* ANSI/ISA-18.2-2009. Research Triangle Park, NC: ISA.

ANSI/ISA. 2011. Wireless systems for industrial automation: Process controls and related applications. 100.11a-2011. Research Triangle Park, NC: ISA.

ANSI/ISA. 2012. *Identification and Mechanical Integrity of Safety Controls, Alarms and Interlocks in the Process Industry.* ANSI/ISA-84.91.01-2012. Research Triangle Park, NC: ISA.

API (American Petroleum Institute). 1991. *Removal of Benzene from Refinery Wastewater.* Standard 221. Washington, DC: API.

API. 1996. *Evaluation of the Design Criteria for Storage Tanks with Frangible Roofs Joints.* Publication 937. Washington, DC: API.

API. 2002. *Pumps – Shaft Sealing Systems for Centrifugal and Rotary Pumps,* 2nd Edition. Standard 682. Washington, DC: API.

API. 2009a. *Tank Inspection, Repair, Alteration, and Reconstruction.* Standard 653. Washington, DC: API.

API. 2009b. *Valve Inspection and Testing.* Standard 598. Washington, DC: API.

API. 2009c. *Venting Atmospheric and Low-Pressure Storage Tanks,* 6th Edition. API 2000/ISO 28300 (Identical). Washington, DC: API.

API. 2013. *Welded Storage Tanks for Oil Storage,* 12th Edition. Standard 650. Washington, DC: API.

ASME (American Society of Mechanical Engineers). 2013. ASME Boiler and Pressure Vessel Code Section VIII – *Rules for Construction of Pressure Vessels,* Division 1. New York: ASME.

ASTM (American Society for Testing and Materials). 2010. *Standard Test Method for Performance Testing of Excess Flow Valves.* Standard F1802-04. Washington, DC: ASTM International.

ATEX (Appareils destinés à être utilisés en Atmosphères Explosives). 2009. *Equipment for Explosive Atmospheres.* Directive 94/9/EC. Brussels: EC.

Bento J.-P., S. Björe, G. Ericsson, A. Hasler, C.-D. Lydén, L. Wallin, K. Pörn, O. Akerlund. 1987. *Reliability Data Book for Components in Swedish Nuclear Power Plants.* RKS/SKI 85-25. Stockholm: Swedish Nuclear Power Inspectorate and Nuclear Training & Safety Center of the Swedish Utilities.

BS (British Standards Institution). 1984. *Specification for Valves for Cryogenic Service.* 6364. London: IBN.

BS. 2006. *Dust explosion venting protective systems.* EN 14491. London: IBN.

Bukowski, J. and W. Goble. 2009. "Analysis of Pressure Relief Valve Proof Test Data." Paper presented at 2009 AIChE Spring Meeting/11th Plant Process Safety Symposium, Tampa, FL, April 26-30.

CCPS. 1989. *Guidelines for Process Equipment Reliability Data, with Data Tables.* New York: AIChE.

CCPS. 1993. *Guidelines for Safe Automation of Chemical Processes.* New York: AIChE.

CCPS. 1998b. *Guidelines for Pressure Relief and Effluent Handling Systems.* New York: AIChE.

CCPS. 2001. *Layer of Protection Analysis: Simplified Process Risk Assessment.* New York: AIChE.

CCPS. 2007b. *Guidelines for Safe and Reliable Instrumented Protective Systems.* New York: AIChE.

CDC (Centers for Disease Control and Prevention). 1998. *Controlling Cleaning-Solvent Vapors at Small Printers.* Number 98-107. Washington, DC: NIOSH.

Chlorine Institute. 2011. *Bulk Storage of Liquid Chlorine,* 8th Edition. Pamphlet 5. Arlington, VA: Chlorine Institute.

Crane Co. 2009. *Flow of Fluids: Through Valves, Fittings and Pipe.* Technical Paper 410. Stamford, CT: Crane Company.

DIN (Deutsches Institut für Normung). 1997. *Plastics Piping Systems – Thermoplastic Valves – Test Methods for Internal Pressure and Leaktightness.* EN 917. Berlin: Beuth Verlag BmbH.

DIN. 2006. *Explosion suppression systems.* EN 14373:2006. Berlin: Beuth Verlag BmbH.

Freeman, R., and D. Shaw. 1988. "Sizing Excess Flow Valves," *Plant/Operations Progress,* July, Vol. 7, No. 3:176-182.

Gertman, D., and H. Blackman. 1994. *Human Reliability and Safety Analysis Data Handbook*. New York: John Wiley & Sons.

Goodrich, M. 2010. "Risk Based Pump-Seal Selection Guideline Complementing ISO21049/API 682." *Proceedings of the Twenty-Sixth International Pump Users Symposium*, Texas A&M University. http://turbolab.tamu.edu/articles/
26th_international_pump_users_symposium_proceedings

Grossel, S. 2002. *Deflagration and Detonation Flame Arresters*. New York: CCPS/AIChE.

Grossel, S. and R. Zalosh. 2005. *Guidelines for Safe Handling of Powders and Bulk Solids*. New York: CCPS/AIChE.

IEC (International Electrotechnical Commission). 2003. *Functional Safety: Safety Instrumented Systems for the Process Industry Sector – Part 1: Framework, Definitions, System, Hardware and Software Requirements*. IEC 61511. Geneva: IEC.

IEC. 2010. *Functional safety of electrical/electronic/programmable electronic safety related systems*. IEC 61508. Geneva: IEC.

ISA (International Society of Automation). 2002. *Safety Instrumented Functions (SIF) – Safety Integrity Level (SIL) Evaluation Techniques*. TR84.00.02-2002. Research Triangle Park, NC: ISA.

ISA. 2011. *Guidelines for the Implementation of ANSI/ISA 84.00.01*. TR84.00.04-2011 Part 1. Research Triangle Park, NC: ISA.

ISA. 2012. *Mechanical Integrity of Safety Instrumented Systems (SIS)*. TR84.00.03-2012. Research Triangle Park, NC: ISA.

ISO (International Organization for Standardization). 1985. *Explosion protection systems -- Part 4: Determination of efficacy of explosion suppression systems*. 6184-4:1985. Geneva, ISO.

MSS (Manufacturers Standardization Society). 2009. *Pressure Testing of Valves*. MSS-SP-61. Vienna, VA: MSS.

Mudan, K. S. 1984. "Thermal Radiation Hazards from Hydrocarbon Pool Fires." *Progress in Energy Combustion Science*. 10:59-81.

NFPA (National Fire Protection Association). 2007. *Standard on Explosion Protection by Deflagration Venting*. 68. Quincy, MA: NFPA.

NFPA. 2008a. *Flammable and Combustible Liquids Code*. 30. Quincy: NFPA.

NFPA. 2008b. *Standard for the Inspection, Testing, and Maintenance of Water-Based Fire Protection Systems*. 25. Quincy, MA: NFPA.

NFPA. 2008c. *Standard on Explosion Prevention Systems*. 69. Quincy, MA: NFPA.

NFPA. 2009. *Standard for Dry Chemical Extinguishing Systems*. 17. Quincy, MA: NFPA.

NFPA. 2010a. *Standard for Exhaust Systems for Air Conveying of Vapors, Gases, Mists, and Noncombustible Particulate Solids*. 91. Quincy, MA: NFPA.

NFPA. 2010b. *Standard for Low-, Medium-, and High-Expansion Foam*. 11. Quincy, MA: NFPA.

NFPA. 2012. *Standard on Clean Agent Fire Extinguishing Systems*. 2001. Quincy, MA: NFPA.

NFPA. 2013. *Standard for the Prevention of Fire and Dust Explosions from the Manufacturing, Processing, and Handling of Combustible Particulate Solids*. 654. Quincy, MA: NFPA.

SINTEF (Stiftelsen for industriell og teknisk forskning). 2010. *PDS Data Handbook*. Trondheim, Norway: SINTEF.

Spirax Sarco. 2013. "Steam Engine Tutorials," Tutorial 12, "Piping Ancillaries," http://www.spiraxsarco.com/resources/steam-engineering-tutorials.asp.

Summers, A. 2014. "Safety Controls, Alarms, and Interlocks as IPLs." *Process Safety Progress*, Vol. 33, No. 2, June 2014, p. 186-194.

Swain, A., and H. Guttmann. 1983. *Handbook of Human Reliability Analysis with Emphasis on Nuclear Power Plant Applications*. NUREG CR-1278. Washington, DC: U.S. Nuclear Regulatory Commission.

UL (Underwriters Laboratories). 2007. *Steel Aboveground Tanks for Flammable and Combustible Liquids*. UL-142. Northbrook, IL: UL.

U.S. CSB (Chemical Safety and Hazard Investigation Board). 2007. *Emergency Shutdown Systems for Chlorine Transfer*. Safety Bulletin No. 2005-06-I-LA. Washington, DC: CSB.

U.S. EPA (Environmental Protection Agency). 2003. *Cross Connection Control Manual*. EPA 816-R-03-002. Washington, DC: EPA.

U.S. EPA. 2007. *Emergency Isolation for Hazardous Material Fluid Transfer Systems – Application and Limitations of Excess Flow Valves*. Chemical Safety Alert, EPA 550-F-0-7001. Washington, DC: EPA.

U.S. EPA. 2009. ECA Workshop for Alabama Correction Facilities, SPCC Review. Washington, DC: EPA.

U.S. OSHA (Occupational Safety & Health Administration). 2005. *Flammable Liquids*. 29 CFR 1910.106. 1974-2005. Washington, DC: OSHA

VDI (Verein Deutscher Ingenieure). 2002. *Pressure Venting of Dust Explosions*. 3673. Berlin: VDI.

6

ADVANCED LOPA TOPICS

6.1   PURPOSE

The LOPA method is intended to be simple and structured, such that the technique achieves consistent results without detailed quantitative analysis. However, there are certain scenarios when the application of LOPA, as presented in the previous chapters, is insufficient. For instance, the conservative assumptions used in LOPA may lead to overestimation of risk and may therefore result in the implementation of more IPLs than necessary. On the other hand, a lack of independence among the IE and IPLs may result in the underestimation of the risk and implementation of fewer IPLs than necessary. In these cases, it may be appropriate to supplement the LOPA with more detailed analysis.

However, it is also worth noting that, although there are some advantages to more quantitative methods, the use of advanced techniques does require a greater degree of knowledge. An understanding of more sophisticated modeling tools and structured logic techniques is necessary to conduct more advanced analyses.

This chapter presents several issues that warrant more detailed analysis along with how they can be addressed. Topics discussed include

- The use of quantitative methods in conjunction with, or instead of, LOPA
- Example application of fault tree analysis (FTA) to develop an appropriate IEF value
- The use of human reliability analysis (HRA) to assess a human IE
- Complex mitigative IPLs

6.2   USE OF QRA METHODS RELATIVE TO LOPA

LOPA is a simplified risk assessment tool that has inherent limitations due to the rules established for the method. There are times when the analyst, site, or company may need to go beyond the simple, limiting rules of LOPA. Alternative quantitative risk assessment methodologies are referenced in Chapter 2.

6.2.1   Use of QRA Methods in Conjunction with LOPA

QRA methods can be used in conjunction with LOPA to help an organization evaluate the PFD of a specific IPL or the specific IEF of an IE. Commonly used QRA methods for assessing likelihood include

- Fault Tree Analysis (FTA) – used to determine the frequency of a defined "top event" by logically combining different system failures, human errors, and process/environmental conditions. The top event is generally the IE frequency, the IPL PFD, or the scenario frequency (which includes the IE frequency and any IPL PFDs).

- Event Tree Analysis (ETA) – follows the logical sequence of potential outcomes as a result of an initiating event. This method explicitly looks at all outcomes that can occur due to the actions of preventive and mitigative IPLs. ETA may also be used to explicitly model the different outcomes related to conditions present after the loss event.

- Human Reliability Analysis (HRA) – a collection of methods and techniques that are available for predicting human error probabilities. These methods include human reliability trees and various simplifications of this approach. These methods include

  - THERP (Technique for Human Error Rate Prediction) (Swain and Guttmann 1983)

  - HEART (Human Error Assessment and Reduction Technique) (Kirwan 1994)

  - SLIM (Success Likelihood Index Method) (Embrey et al. 1984)

  - HCR (Human Cognitive Reliability) (Hannaman, Spurgin, and Lukic 1984)

  - APJ (Absolute Probability Judgment) (Kirwan 1994)

  - SPAR-H (Standardized Plant Analysis Risk-Human Reliability Analysis method) (Gertman et al. 2005)

In addition to quantitative methods for assessing the likelihood of an event, there are also quantitative methods for assessing the potential consequence of an event. General modeling of fire, explosion, and toxic impacts can be done using public software such as Areal Locations of Hazardous Atmospheres (ALOHA) (NOAA/U.S. EPA 2012), which is often sufficient for LOPA. For detailed modeling, it may be necessary to use more sophisticated, proprietary software.

QRA methods are described in detail in other references, such as *Guidelines for Chemical Process Quantitative Risk Analysis*, 2nd Edition (CCPS 2000). One of the best publications on HRA is the *Handbook of Human Reliability Analysis with Emphasis on Nuclear Power Plant Applications, Final Report*, NUREG CR-1278 (Swain and Guttmann 1983). A good reference that provides a method for simplified human error estimation is *The SPAR-H Human Reliability Analysis Method*, NUREG RC-6883 (Gertman et al. 2005).

6.2.2   Use of QRA Methods Instead of LOPA

The use of more quantitative risk assessment methods may require additional analyst training and a significant investment of time to complete the quantitative assessment. However, there are situations when an organization may wish to make this investment to move beyond LOPA to QRA. Some organizations do the following:

- Utilize QRA methods if either the consequence or the frequency of the scenario is not well understood by the LOPA analyst or LOPA team.

- Require use of QRA to validate the required SIL of a SIF.

- Use QRA to evaluate scenarios that have safeguards with significant common-cause failures. The simplified rules of LOPA do not generally allow credit for IPLs that are not fully independent; this can sometimes result in an overestimation of the risk of the event.

- Apply a QRA methodology if the potential consequence severity is very large, such as a consequence that could potentially result in a high number of fatalities or create a catastrophic impact on the environment.

- Move beyond LOPA if the results are on the borderline of tolerable risk. To avoid potentially expending a significant amount of capital on unnecessary risk reduction, a site may instead choose to perform a QRA to determine whether a more rigorous approach indicates that the risk is actually at a tolerable level. Alternatively, further analysis using QRA may instead indicate that additional risk reduction is required to achieve the organization's criteria for tolerable risk.

6.2.3   Example: FTA to Evaluate a Complex IE

In LOPA, the analyst evaluates each scenario and estimates the IE frequency, such as a failure of a control loop giving rise to high temperature with an IE frequency of 0.1/yr. Some organizations apply LOPA to evaluate defined major events. This determination can be difficult to perform using LOPA, because of the number of causes and the potential for these causes to have shared equipment, systems, personnel and management systems that need to be considered. Predictive techniques like FTA may be used to estimate the IE frequency based upon the design, operation, and maintenance of the systems that lead to the major event. The fault tree approach is particularly valuable because it can address complex systems that violate the strict LOPA rules for independence (e.g., primary and secondary booster pumps that share a common power supply or valving).

FTA combines the causes of individual scenarios in a logical but parallel order. The events are linked by 'OR' and/or 'AND' gates, depending on their relationship. Fault trees can be extremely complex; however, the level of detail and complexity of the analysis can be weighed against the requirements of the analysis.

A simple fault tree example is given in Figure 6.1 to illustrate the evaluation of multiple causes to determine an overall IE frequency for a major event, high temperature in a reactor vessel. Figure 6.1 provides a basic demonstration of fault tree logic and math.

In this example, high temperature is caused either by cooling water failure or by the incorrect charging of the reactants or catalyst. The illustration shows a simple, high level fault tree. Typically, each cause of the high temperature IE is analyzed in more detail to identify the specific equipment, system, and operator failures that can lead to the event. For example, too much catalyst being charged may be caused by catalyst flow control loop failure, operator error, or other identified failures

The following assumes that the analyst has already determined the frequency of the events illustrated and that the events are sufficiently independent from one another such that the simple math shown below is appropriate. It also assumes that any IPLs that are considered in the LOPA are capable of addressing all of the causes considered in the IE frequency analysis.

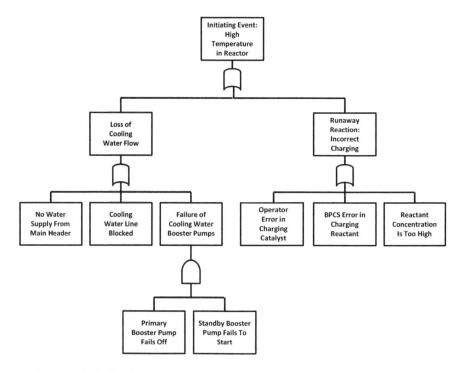

Figure 6.1. Fault tree for example of high temperature in the reactor.

The *loss of cooling water flow* is caused either by the lack of cooling water supply from the main header, the cooling water line being blocked, or the cooling water pumps failing. For the failure of the cooling water pumps, both the electric main pump and the diesel standby pump fail. In fault tree math, the failure rates of the individual causes that go through the OR gates are added to derive the composite failure rate. (For AND gates, the values of the subevents are multiplied to calculate a composite value for a particular event.) Therefore, given the failure rates for the following three basic causes of loss of cooling:

- No supply of cooling water from its source has occurred approximately every 10 years (IEF = 0.10/yr)
- Line blockage has occurred once in 20 years (IEF = 0.05/yr)
- Cooling water pump failure has occurred once in 10 years (IEF = 0.10/yr) AND the probability of failure of standby pump = 0.1 per demand. Therefore, the composite failure rate for the cooling water pumps system is 0.1/yr x 0.1 = 0.01/yr.

The IEF for loss of cooling water flow can be approximated by

$IEF_{loss\ of\ cooling\ water\ flow}$ = 0.10/yr + 0.05/yr + 0.01/yr = **0.16/yr**.

For the same fault tree, the *incorrect charging of the feed* to the reactor can either be caused by the operator charging too much catalyst, the BPCS adding an incorrect amount of the reactant, or an incorrect concentration of reactant being supplied to the facility. Assuming the following failure rate data:

*   There are 25 charges per year, so the frequency of operator error in charging catalyst or reactant = 0.1/yr per Data Table 4.4.
*   The frequency of the BPCS failure in charging the reactant = 0.1/yr per Data Table 4.1.
*   The frequency that the reactant is supplied at the wrong concentration is estimated to be = 0.05/yr.

Then the frequency of the *incorrect charging of the feed* is

$IEF_{incorrect\ charging}$ = 0.1/yr + 0.1/yr + 0.05/yr = **0.25/yr**.

Since either *loss of cooling water flow* or *incorrect charging of the feed* can cause high temperature in the reactor, the combined frequency of high temperature in the reactor would be

$IEF_{high\ temperature\ in\ reactor}$ = 0.16/yr + 0.25/yr = **0.41/yr**.

If this value were to be used in order of magnitude LOPA, the IEF of high temperature in the reactor would be rounded up to 1/yr.

This is only a brief example of fault tree structure and math. A comprehensive introduction to QRA techniques, including fault tree analysis, can be found in *Guidelines for Chemical Process Quantitative Risk Assessment* (CCPS 2000).

6.2.4   Use of HRA to Evaluate a Human IE

A common HRA tool for complex human actions is the HRA event tree, which is associated with the THERP method cited above (Swain and Guttmann 1983). It incorporates a pyramid of branches representing human success and failure paths. Paths branching to the left are usually considered as the success routes, while branches to the right are failure paths. The branches terminate at a point when the task is successfully completed or when an unrecoverable error occurs. The tree can be solved mathematically using conditional probability, where the probability of the successful completion of a task or step depends on the success or failure of the previous step or task. This is represented in Figure 6.2.

Definitions of items in the example HRA tree:

    Task "A" = the first task

    Task "B" = the second task

    a = probability of the successful performance of Task A

    A = probability of the unsuccessful performance of Task A

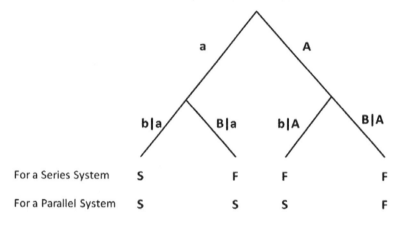

Figure 6.2. Human reliability analysis (HRA) tree logic (Lorenzo, 1990).

    b|a = Probability of the successful performance of Task B, given a

    B|a = Probability of the unsuccessful performance of Task B, given a

    b|A = Probability of the successful performance of Task B, given A

    B|A = Probability of the unsuccessful performance of Task B, given A

For a Series System:

    Probability of success (S) = a x (b|a)

    Probability of failure (F) = 1 - a x (b|a) = a x (B|a) + A x (b|A) +
                                            A x (B|A)
                                     = a x (B|a) + A

For a Parallel System:

    Probability of success (S) = 1 - A x (B|A)= a x (b|a) + a x (B|a) +
                                              A x (b|A)
                                       = a + A x (b|A)

Probability of failure (F) = A x (B|A)

Appendix A provides more information on human factors and their effects on human reliability. However, an analyst generally needs training and experience to become proficient in the application of quantitative HRA methods.

6.3   EVALUATION OF COMPLEX MITIGATIVE IPLS

As stated in Chapter 5, an IPL is a device, system, or action that is capable of preventing a scenario from developing to the undesired consequences, independent of the IE or any other IPL associated with the scenario. IPLs used to reduce the frequency of realizing the consequence of concern after a loss of containment are termed mitigative systems or mitigative IPLs. Mitigative IPLs reduce the frequency of progression to the original, more severe, consequence but may allow, or even facilitate, less severe consequences to occur. For instance, a relief valve prevents catastrophic failure of a pressure vessel but, in doing so, acts as a conduit to the atmosphere that may produce another consequence of concern. Most mitigative IPLs are best evaluated in two scenarios, one with success of the mitigative IPL and one with failure of the mitigative IPL.

Some mitigative IPLs, such as dikes, berms, and bunds, are relatively straightforward, and it is possible to develop generic PFD values for them. Other mitigative IPLs are application-specific and quite complex. Evaluating a complex mitigative IPL to obtain an appropriate PFD can be challenging because a number of factors can influence the likelihood (probability) that the IPL would be effective. Determining an appropriate PFD value for complex IPLs may require the use of more quantitative analysis methods than LOPA, such as those discussed in this chapter. Some examples of complex mitigative safeguards are given below.

Abatement Systems

Abatement systems are used to remove residual hazardous chemicals from process streams by various methods, such as neutralization, absorption, combustion, etc. These systems may also be used to manage larger releases of material, such as during emergency venting of a toxic or flammable material. An abatement system generally has a finite capacity; this should be considered when determining an appropriate IPL value. If the quantity, discharge rate, or concentration during the event exceeds the system capacity, the abatement system may not be an effective IPL.

Abatement systems may be composed of several mechanical and instrumented systems that function together. Other systems may have fewer active components, such as flares or sparge tanks. The IPL PFD of each component of an abatement system is considered when selecting a value to be used for an overall PFD. Determination of the PFD for an abatement system is complex and cannot be represented by a generic LOPA PFD value. More detailed methods, such as those discussed in this chapter, may be required to determine the overall PFD for an abatement system.

Access Control Safeguards

In some situations, access control safeguards are used to restrict entry to areas during the performance of higher-risk activities, such as during maintenance or process startups. Access control safeguards do not act to prevent a hazard scenario from occurring; rather, they act to limit the number of people affected by restricting the occupancy in the area of potential impact. Access control safeguards can include the use of barricades, signs, and barrier tape. These types of safeguards can be effective when used within a strong culture of operating discipline. However, since these safeguards are specific to a site and to a particular situation, it may be appropriate to use more quantitative risk assessment methods to assess their effectiveness. Alternatively, some companies consider access control as a personnel-presence conditional modifier rather than an IPL. Additional information regarding the use of conditional modifiers can be found in *Guidelines for Enabling Conditions and Conditional Modifiers in Layer of Protection Analysis* (CCPS 2013).

Emergency Response Safeguards

An effective emergency response plan, and the availability of emergency facilities within an installation, can significantly reduce the number of casualties, environmental harm, and damage to equipment as a result of an event. An emergency plan that documents the response to a major incident for the purpose of mitigating the impact of the event is an important safeguard for a facility. However, it can be challenging to evaluate the effectiveness of an emergency response safeguard. Some examples of emergency response safeguards are:

- *Firefighting:* Manual firefighting activities have proven to be very valuable in emergency response. However, it is not possible to develop a generic IPL value reflecting the effectiveness of such activities. The equipment available to fight the fire, the response time, the training of individuals, and the likelihood of putting out a particular type of fire is situation-specific. During an emergency, there are other factors, such as stress, that can adversely affect the response activity.

- *Emergency Response and Evacuation:* As with firefighting, emergency response and evacuation can be very effective in preventing serious injury and fatality. These safeguards are most effective when there is a long enough time between the detection of the abnormal condition and the onset of the consequence, which allows for effective emergency response or for evacuation of the nearby community. Under certain circumstances, such activities can be an effective means of limiting harm, but assessment of the effectiveness of such safeguards would require analysis using more quantitative risk assessment methodologies.

6.4 CONCLUSIONS

LOPA is a useful, streamlined method for assessing risk. There may, however, be times when LOPA might not be the best method to assess the risk of a particular scenario and more quantitative risk assessment methods may be required. There may also be times when quantitative risk assessment methods can be effectively used to supplement a LOPA analysis. This chapter discussed some of the available quantitative risk assessment methodologies, including the application of fault tree analysis (FTA) and human reliability assessment (HRA). Examples of complex mitigative safeguards were also provided; the use of quantitative risk assessment methods may be helpful to determine if safeguards such as these could be effective enough to potentially qualify as IPLs.

REFERENCES

ANSI/API (American National Standards Institute/American Petroleum Institute). 2008. *Guide for Pressure-Relieving and Depressuring Systems: Petroleum petrochemical and natural gas industries-Pressure relieving and depressuring systems,* 5th Edition addendum. (ISO 23251 Identical). API Standard 521. Washington, DC: API.

CCPS. 2000. *Guidelines for Chemical Process Quantitative Risk Analysis,* 2nd Edition. New York: AIChE.

CCPS. 2013. *Guidelines for Enabling Conditions and Conditional Modifiers in Layer of Protection Analysis.* New York: AIChE.

Embrey, D., P. Humphreys, E. Rosa, B. Kirwan, & K. Rea. 1984. *SLIM-MAUD: An approach to assessing human error probabilities using structured expert judgement.* NUREG/CR-3518. Washington, DC: U.S. Nuclear Regulatory Commission.

Gertman, D., H. Blackman, J. Marble, J. Byers, and C. Smith. 2005. *The SPAR-H Human Reliability Analysis Method.* NUREG CR-6883. Washington, DC: U.S. Nuclear Regulatory Commission, Office of Nuclear Regulatory Research.

Hannaman, G., A. Spurgin, and Y. Lukic. 1984. *Human cognitive reliability model for PRA analysis.* Draft Report NUS-4531, EPRI Project RP2170-3. Palo Alto, CA: Electric Power and Research Institute.

Kirwan, B. 1994. *A Guide to Practical Human Reliability Assessment.* Boca Raton, FL: CRC Press.

Lorenzo, D. 1990. *A Manager's Guide to Reducing Human Error: Improving Human Performance in the Chemical Industry.* Washington, DC: Chemical Manufacturers Association.

NOAA/U.S. EPA (National Oceanic and Atmospheric Administration/U.S. Environmental Protection Agency). 2012. ALOHA (Areal Locations of Hazardous Atmospheres) (Version 5.4.3) [computer program] http://www.epa.gov/emergencies/content/cameo/aloha.htm.

Swain, A., and H. Guttmann. 1983. *Handbook of Human Reliability Analysis with Emphasis on Nuclear Power Plant Applications*. NUREG CR-1278. Washington, DC: U.S. Nuclear Regulatory Commission.

APPENDICES

APPENDIX A. HUMAN FACTORS CONSIDERATIONS

INTRODUCTION

Humans can be the cause of an incident scenario, and humans can also be part of an independent layer protecting against an incident scenario. Other CCPS publications, such as *Guidelines for Preventing Human Error in Process Safety* (CCPS 1994) and *Human Factors Methods for Improving Performance in the Process Industries* (CCPS 2007c), comprehensively address human factors issues. This appendix serves to highlight key areas that influence the selection of IE and IPL values and affect the applicability of the generic values presented in this document.

Although human error comes in many forms, the primary human errors of interest in LOPA are those that directly relate to human IEs and IPLs. These include

- Errors in following procedures, such as the startup of a unit, that initiate a LOPA scenario
- Errors in responding to a call for action that, if otherwise performed correctly, would have prevented the consequence from occurring

It is important to understand the various factors that influence human error so that they can be effectively managed to support high levels of human performance.

WHAT IS HUMAN ERROR?

J. Reason (1990) offered the following definition when discussing human error:

> *"Error will be taken as a generic term to encompass all those occasions in which a planned sequence of mental or physical activities fails to achieve its intended outcome..."*

Human error is one of the major contributors to personal injury, property damage, and environmental impact within the chemical industry. In the older, traditional viewpoint, human error was considered to be the "root cause" of incidents. The traditional approach often involved a culture of blame and punishment and failed to recognize the contribution of management system failures to human error.

The systematic viewpoint on human error focuses not on the failings of the individual but rather on the conditions that created an environment where the error could occur. *Guidelines for Preventing Human Error in Process Safety* (CCPS 1994) viewed human error as

"a natural consequence of a mismatch between human capabilities and demands, and an inappropriate organizational culture."

When viewed within this framework, to achieve a high level of human performance, it is necessary to understand not only the types of human errors that can occur but also the management systems that influence them. The key elements within these management systems can then be appropriately controlled to create a higher level of reliability within the organization and a lower error rate among individuals.

CATEGORIZATION OF HUMAN ERRORS

Over the past several decades, a number of different models have been offered as frameworks within which one can assess human error. Understanding the types of errors that can occur, and the management systems that can influence each type of error, is useful in identifying specific factors that can be controlled to ensure sufficiently reliable human performance.

Omission/Commission/Sequence/Time

Swain and Guttmann (1983) proposed four categories of human error with respect to operator actions:

1. Error of Omission – when an operator fails to perform a necessary step or task.
2. Error of Commission – when an operator performs a step incorrectly or performs an act other than the correct step.
3. Sequential Error – when an action is performed out of the correct order.
4. Time Error – when an action is performed too early, too late, too slowly, or too quickly.

These error descriptions can be useful to PHA teams to facilitate discussions of what types of errors can occur in a given process and what human error initiating events and scenarios can be considered in LOPA. However, the categories describe *what* occurred rather than *why* it occurred, and the descriptions do not help to identify what gaps in management systems might have impacted the probability of an error. It is important to understand which human factors are most important to control to achieve the desired human performance for a particular IE or IPL.

Skill-Based/Rule-Based/Knowledge-Based Mistakes

Rasmussen and Rouse (1981) referred to three basic types of operator action. These types are differentiated by increasing levels of conscious operator attention that are needed to perform the required tasks.

1. Skill-based: Repetitive actions requiring little conscious thought (e.g., driving a forklift truck or reading a gauge).
2. Rule-based: Actions specified by some clear directive, such as a verbal instruction or a written procedure.
3. Knowledge-based: Situations requiring conscious thought about what action to take (e.g., diagnosing a process upset).

Skill-based tasks are performed routinely and require little conscious thought. Skill-based errors occur when the wrong action is performed or the correct action is omitted. The frequency of skill-based errors can be reduced by providing routine practice and by driving strong work habits in a culture that does not tolerate "short cuts." Physical limits, such as hose fittings with different sizes or thread pitch, can significantly reduce the likelihood of error on skill-based tasks. It is also important for an organization to be cognizant when operational changes are driving corresponding changes to longstanding work practices, so that a formerly appropriate routine action does not become a human error.

Rule-based errors occur when the wrong rule or action is selected for an action, or the correct rule is executed incorrectly. Rule-based activities generally require more conscious thought than skill-based activities. Rule-based error rates can be reduced by having accurate, well-written procedures and checklists available to the individual, and by providing appropriate initial and refresher training. It is also important to have a culture that reinforces the need to follow rules and procedures rigorously.

Knowledge-based errors occur when the wrong plan is generated for a given situation. These situations often arise when there is no established procedure for responding to a situation, and the human is required to determine an appropriate response. These situations require the most mental processing and rely on a higher level of knowledge than skill- or rule-based tasks. The probability of a knowledge-based mistake can be reduced by developing symptom-based procedures and troubleshooting guides. Symptom-based procedures are instructions that are to be followed for a specific process deviation (or "symptom"), without requiring the operator to be able to identify the specific cause of the deviation. (One analogy is when a doctor orders a cool water bath for a patient with a high fever without knowing the specific cause of the fever.) Knowledge-based mistakes can also be reduced by assigning people who are capable of handling the higher level of mental processing required in these situations. Training and experience are the most common means by which individuals gain the skills necessary to perform knowledge-based tasks.

This categorization of human errors is helpful in that it gives some indication of which management systems can most influence the human error probability for a particular type of human error. The next section discusses in more detail the individual factors that influence human performance, and the degree to which they affect the probability of human error.

PERFORMANCE SHAPING FACTORS

Every risk assessment needs to consider the likelihood of human error and the effect of human factors on the human IE or IPL PFD value selected. Poor control of human factors can lead to higher IE frequencies and higher PFDs for human IPLs. Further, failure to sufficiently control human factors may invalidate the use of the generic values provided in the data tables in this document.

Considerations Regarding Performance Shaping Factors

Performance shaping factors (PSFs) are influences that enhance or degrade human performance. They can be described as those factors that determine the probability of error or the level of human performance relative to the situation that would exist in the absence of the PSFs. PSFs include conditions in the work environment, job/task instructions, task and equipment characteristics, psychological stressors, physical stressors, and characteristics of the individual human performing the task. The degree to which PSFs are favorable or unfavorable is ultimately determined by an organization's management systems.

Much of the data that exist on the numerical impact of PSFs on the likelihood of human error have come from the nuclear industry. Some data have been collected in the chemical process industries (e.g., Edwards and Lees 1974). Collecting data on human error is challenging because errors may not be frequent or visible. Workers are typically reluctant to report errors that are caught and corrected before harm occurs due to fear of repercussions. Moreover, in many cases, there is no systematic way to record, capture, and analyze the data. It is frequently necessary to rely on expert judgment where hard data do not exist.

IMPACT OF PERFORMANCE SHAPING FACTORS ON HUMAN ERROR PROBABILITIES

Data on the impact of PSFs on human performance are frequently expressed in terms of human error probability (HEP). *Guidelines for Preventing Human Error in Process Safety* (CCPS 1994) defined HEP as

> *"The probability that an error will occur during the performance of a particular job or task within a defined time period.*

> Alternative definition: *The probability that the human operator will fail to provide the required system function within the required time."*

The intent of this section is to provide the reader with a perspective on the degree to which human error probability can be impacted by PSFs. It is hoped that the reader will better appreciate the assumptions underlying the generic IEF and IPL values provided in the data tables within this document. However, the reader is cautioned against directly applying values contained in the subsequent tables, except to increase generic human IE or IPL values for LOPA. For example, displays violating strong group expectations (also called "populational stereotypes") increase the likelihood of human error by a factor of 10; the nominal values in this document assume no such gross violations exist. Before applying any data, it is important to understand the source of the data and the assumptions inherent in the data to determine if the data are truly applicable for a specific situation.

Complexity

HEPs are impacted by the degree of complexity of an activity. HEP tends to be lower when procedures are simple, step-by-step, and do not include calculations or interpretation. HEP tends to be higher when procedures are complex, such as when there are many steps, the procedures include calculations, and/or completion of the procedure requires judgment. For example, according to Swain and Guttmann (1983), a checklist containing more than 10 steps can have a HEP that is a factor of 3 higher than that of a shorter checklist.

Checklists and Procedures

Workers following written procedures and/or checklists have been shown to have lower HEPs than those working from memory. In the *Handbook of Human Reliability Analysis with Emphasis on Nuclear Power Plant Applications, Final Report,* NUREG CR-1278 (Swain and Guttmann 1983), error rates were tabulated from various sources, including some empirical data. Errors that could be considered initiating events were summarized in Table 15-3 of that document, which is extracted and included in slightly amended form below in Table A-1. Note, this table applies only for errors of omission (skipping a step), not for errors of commission (doing the step incorrectly).

Table A-1. Baseline human error rate for initiating event type of tasks (extracted from NUREG CR-1278 [Swain and Guttmann 1983]; Table 15-3)

Omission of Item	Human Error Probability
When procedures with check-off provisions are correctly used	
(1) Short list, less than or equal to 10 items	0.001
(2) Long list, more than 10 items	0.003
When procedures without check-off provisions are used, or when available check-off provisions are incorrectly used	
(3) Short list, less than or equal to 10 items	0.003
(4) Long list, more than 10 items	0.01
(5) When written procedures are available and should be used but are not used	0.05

Swain and Guttmann's data (1983) indicated that the correct use of a checklist (with the list physically in hand and items individually checked off as work is completed) can reduce the HEP by a factor of 3. However, if written procedures are required but not used, the HEP can increase by a factor of 20 to 50, based on the referenced information. Clearly, procedures and checklists can be effective tools in reducing the likelihood of human error, especially for rule-based mistakes.

Ergonomics

Ergonomics involves designing the interaction between the human and the environment to be conducive to effective performance. Some specific PSFs to be considered include

1. Human Factors Engineering – Examples of factors to evaluate when designing the interface between the operator and the process include

    - Design and layout of controls and displays
    - Alarm management to avoid excessive or nuisance alarms
    - Indications of equipment position or status that are consistent with local conventions
    - Clear, understandable labeling
    - Effective communications practices

    It is important to consider the cultural norms of the individuals who will be operating the equipment. Strong populational stereotypes are expectations about the meaning of a signal or the operation of a device that are reinforced by everyday experience. Stopping a car at a red traffic light or turning a handle clockwise to close a valve are examples of strong populational stereotypes. Violating these stereotypes can increase the frequency of human error. In some cultures, a pump shown in red on a computer display may mean "off" to an operator; in other cultures, red may mean "energized." It is important to consider such cultural norms when designing displays and work practices to decrease human error rates.

2. Work Environment – Human performance is influenced by factors such as

   - Lighting
   - Temperature
   - Noise

3. Individual Stressors – Physical and psychological factors can influence the ability of the individual to perform the desired activity correctly; they include

   - Level of physical exertion required to perform the action
   - Capability of a human to physically perform the task required

The impact of ergonomic considerations on human error is illustrated in Table A-2, extracted from *Handbook of Human Reliability Analysis with Emphasis on Nuclear Power Plant Applications, Final Report*, NUREG CR-1278 (Swain and Guttmann 1983) Table 3-8.

Table A-2. Impact of ergonomic considerations on human error (extracted from NUREG CR-1278 [Swain and Guttmann 1983] Table 3-8)

If Done	Resulting Decrease in HEPs (Factors*)
Good human factors engineering practices in design of controls and displays	2 to 10
Redesign of displays or controls that violate strong populational stereotypes	>10
Redesign of valve labeling to indicate its function (including a clear indication of the system with which a valve is associated) and also to show clearly its normal operating status	~ 5

* These estimated factors are not additive.

The table indicates that good engineering practices, which effectively consider human factors, can decrease the HEP by 2x. Clear labeling of equipment can decrease the HEP by a factor of 5. There is also a significant impact on HEP when local conventions are not considered. Display designs that violate strong populational stereotypes can increase the HEP by 10x or more.

Skill Level and Training

Humans are most effective when they have the skills and training necessary to be effective. In particular, knowledge-based error rates can be reduced by an increased emphasis on training and skill development. Operators require effective initial training as well as periodic refresher training to maintain skill levels. Training methods can include a review of procedures, classroom training, on-the-job training, and computer training. Proficiency can be determined by testing and by field demonstration of the required tasks.

For nonroutine tasks, such as procedures used in the performance of an IPL, procedures and checklists can reduce human error. Other means of increasing performance include field drills, tabletop exercises, and computer simulation. In fact, according to Swain and Guttmann (1983), frequent practice of the appropriate responses to alarm situations can reduce the HEP by a factor of 3–10.

Task Load

The task load represents the performance expectation of an operator with regard to all activities expected of the individual. When the task load is high, the operator is expected to perform at or above his or her capacity. People operating under a heavy task load can experience degradation of performance; however, the degree to which performance deteriorates is difficult to quantify.

Stress and Fatigue

NUREG CR-1278 (Swain and Guttmann 1983) defines stress as

> "a continuum, ranging from a minimal state of arousal to a feeling of threat to one's well-being, requiring action."

Stress can be categorized as

1. Very low (insufficient stimulus to keep alert)
2. Optimum (the facilitative level)
3. Moderately high (slightly to moderately disruptive)
4. Extremely high (very disruptive)

The belief that very low stress improves human reliability is a common misperception. In fact, people need some level of stress to maintain interest and focus. If the stress level is too low, the level of alertness can drop, increasing the probability of human error. This can be an issue for roles such as night security guards or fire watches. There is an optimum level of stress that facilitates higher levels of human performance. When stress is increased beyond the optimum level, it can be disruptive to the individual, and human performance will generally degrade. Table A-3 illustrates the impact of stress on worker performance.

Table A-3. Modification of estimated HEPs for the effects of stress on skilled personnel (extracted from NUREG CR-1278 [Swain and Guttmann 1983] Table 18-1)

Stress Level	Modifiers for Nominal HEPs	
	Skilled *	Novice **
Very low (Very low task load)	x2	x2
Optimum (Optimum task load):		
Step by step +	x1	x1
Dynamic +	x1	x2
Moderately high (Heavy task load):		
Step by step +	x2	x4
Dynamic +	x5	x10
Extremely high (Threat stress):		
Step by step +	x5	x10
Dynamic + Diagnosis ++	0.25 This is the actual HEP to use with dynamic tasks or diagnosis – this is *not* a modifier.	0.50 This is the actual HEP to use with dynamic tasks or diagnosis – this is *not* a modifier.

* A skilled person is one with 6 months or more experience in the tasks being assessed. Even a very senior worker might not qualify as "skilled" in unit startup tasks that are only performed every 3–5 years.

**A novice is a person with less than 6 months of experience. Both levels have the required licensing or certificates (i.e., are fully qualified).

+ Step-by-step tasks are routine, rule-based tasks, such as carrying out written calibration procedures. Dynamic tasks are knowledge-based tasks that require a higher degree of human-machine interaction, such as decision-making, keeping track of several functions, controlling several functions, or any combination of these. These requirements are the basis of the distinction between step-by-step tasks and dynamic tasks, which are often involved in responding to an abnormal event.

++ Diagnosis may be carried out under varying degrees of stress, ranging from optimum to extremely high (threat stress).

Most HEP values presented in literature assume that the individual is under an optimum state of stress. When using literature values, care should be taken in their use if a human involved in a scenario of concern is expected to be operating under very low or very high stress levels. The impact of stress levels on the performance of novice and experienced individuals based on task load can be seen in NUREG CR-1278 (Swain and Guttmann 1983) Table 18-1.

NUREG CR-1278 (Swain and Guttmann 1983) Table 18-1 also suggests that heavy task loads can increase the HEP by a factor of 2–5, depending on whether the activity is a simple, step-by-step activity or an activity involving higher levels of mental processing and interpretation. For novices, high stress levels and task loads can increase the HEP by 10x or more.

The LOPA values relating to human performance assume reasonable control of the workers' fitness for duty, including management of performance shaping factors such as fatigue. The impact of fatigue can result in significantly higher human error rates than assumed in this document. Worker fatigue has been linked to a number of serious industrial events, including the Texas City isomerization unit explosion (The BP U.S. Refineries Independent Safety Review Panel [Baker et al. 2007]) and the Exxon Valdez oil spill (NTSB 1990). High physical and cognitive demands, as well as factors such as sleep deficit, illness, and the use of certain medications, can all contribute to degrading a worker's fitness for duty. Degraded fitness for duty can increase the probability of a human error by up to a factor of 5 (*Human Error Quantification Using Performance Shaping Factors in the SPAR-H Method* [Blackman, Gertman, and Boring 2008]). A good reference on fatigue management is *Fitness for Duty Programs* (U.S. NRC 2008).

Example Application

HEPs can be used to adjust the suggested values provided in these data tables to account for the influence of specific performance shaping factors at a site. For example, although a plant may generally use valves that are turned clockwise to close, a particular task requires a worker to operate a valve that must be turned counter-clockwise to close. This violates a strong group expectation (populational stereotype). Referencing Table A-2, the HEP of the task may be 10x higher due to the influence this PSF. Data Table 4.4 suggests a human error rate of 0.1/yr for a task performed at a frequency of once a week to once a month, assuming that human factors are reasonably controlled. Given the violation of the strong group expectation, the suggested IEF would be adjusted to 0.1/yr x 10 = 1/yr.

DEPENDENCE

Just as fault tree analysis (FTA) can be used to evaluate situations where equipment-related safeguards are not completely independent, human reliability analyses (HRAs) are sometimes used to evaluate situations where human actions (by the same or multiple individuals) may not be completely independent.

Examples of this application of HRA include determining the probability that an operator will fail to identify an error in another operator's work or that an operator will fail to identify a previous error that the operator made and recover from that error before a consequence of concern occurs.

In an HRA, two actions are independent when the probability of performing one action incorrectly is the same whether or not a prior action has been performed incorrectly. In situations where the probability of error in completing a task is influenced by the probability of error in completing a separate, prior task, the second task is considered to be dependent on the first.

The concept of dependency applies between groups of people as well as by the same person. For example, if an experienced reliable operator has failed to close Valve 1, a supervisor would be more likely to fail to detect the error than if the same valve had been left open by a trainee. The supervisor's attention to detail depends on the person being checked. Similarly, if a technician has miscalibrated Instrument 1, he or she would be more likely to miscalibrate Instrument 2 using the same methods and tools on the same day.

Swain and Guttmann (1983) indicate that methods for assessing the degree of dependence include

1. The use of actual data on the operations in question
2. Direct estimation made by qualified personnel based on information from similar operations and the relevant performance shaping factors
3. The use of dependence models

The best method to assess dependence is to use actual data from the specific task being evaluated. If such data are not available, judgmental assessments can be made based on the nature of the task for which there are data and any differences in performance shaping factors that are applicable to the task of interest.

A third approach is the use of models. One such model is Swain and Guttmann's positive dependence model (1983) that estimates the degree of dependence in interactions between human tasks based on a consideration of human factors. The model represents the level of dependence in five categories:

*   Zero
*   Low
*   Medium
*   High
*   Complete

Zero dependence occurs when the performance (or nonperformance) of a task has no impact on the performance (or nonperformance) of a subsequent task. Full independence between two humans performing a task is uncommon. However, if the second person is making independent measurements, and his personal safety depends on the outcome, the second person is usually judged to act independently of the first. For example, a mechanic is usually considered to act independently when verifying that equipment has been properly isolated and de-energized by an operator. The actions are taken by different individuals, at different times (and possibly in different locations), and the personal safety of the maintenance worker is at risk if the operator did not properly isolate the equipment.

Low dependence occurs when an operator tends not to trust or rely on another operator and will likely perform the tasks independently. Experienced co-workers on different shifts often exhibit low dependence on each other, especially if there is a high degree of operating discipline.

Medium dependence occurs when there is an obvious relationship between the performance of one task and the performance of a subsequent task. For example, if an operator has misread a gauge, he is fairly likely to misread a similar gauge.

High dependence occurs when the performance of one task very significantly affects performance of a subsequent task. For example, when one individual has power or authority over another individual, tasks that they perform are likely to have a high level of dependence since the subordinate person is likely to defer to the person of higher rank.

Complete dependence occurs when a second operator does not check his or her own work or the work of another operator, but instead assumes that the work of the first individual was correct. Swain and Guttmann (1983) indicated that instances of complete dependence are uncommon under normal operating circumstances but do occur frequently enough to warrant consideration. In emergency situations, junior workers often simply follow the instructions of senior, more experienced co-workers or supervisors, even if they believe the instructions are incorrect. Likewise, if a worker has gone to the wrong train of equipment and opened Valve 1B instead of 1A, the worker will almost certainly open Valve 2B instead of 2A as well.

The impact of the level of dependence on the HEP is shown in Table A-4.

Table A-4. Human error probability as a function of dependency between operators (extracted from NUREG CR-2254 [Bell and Swain 1983])

Level of Dependence	Human Error Probability for a Second Operator
Zero dependence	Basic HEP for an individual operator
Low dependence	0.05
Moderate dependence	0.15
High dependence	0.5
Complete dependence	1.0

There are no firm guidelines on the evaluation of the level of dependence between operators; it is highly situation-specific and should be carefully assessed by analyzing past error rate data, the dynamics of the interaction between individuals, and the applicable performance shaping factors. This is particularly true if the "second person" is, in fact, the same person performing a different task intended to reveal an error. For example, a worker performing a pressure check to test the integrity of a hose connection might be judged to have low dependence on his or her previous error of incorrectly installing the gasket. However, simply telling a worker to "verify the hose is properly connected" would typically be judged to have moderate to complete dependence on the original gasket installation error. As stated by Swain and Guttmann in NUREG CR-1278 (1983),

> "A general guideline is to assume dependence between human activities unless a diligent search reveals no significant interactions between them."

Thus, for LOPA, the second-person check could potentially qualify as an IPL only if there is low dependence between the worker(s), stress levels are optimum to moderate, and other performance shaping factors are typical for a modern process plant. (Of course, the other core attributes of an IPL still apply. Refer to Chapter 3 for a review of the core attributes of an IPL.) If these conditions are met, a value of 0.1 may be used as the PFD for a second-person check.

SUMMARY: PERFORMANCE SHAPING FACTORS

A worker error is rarely a root cause of an incident; rather, it is usually a symptom of imperfect management systems. As discussed in this section, gaps in systems such as process design, workplace environment, training, task load, and individual stress can have a significant impact on the HEP. As stated in *Managing Maintenance Error: A Practical Guide* (Reason and Hobbs 2003),

> "Errors are consequences, not just causes. They are shaped by local circumstances: by the task, the tools and equipment and the workplace in general. If we are to understand the significance of these factors, we have to stand back from what went on in the error maker's head and consider the nature of the system as a whole."

HUMAN ERROR RATE AND INITIATING EVENT FREQUENCY

The frequency of human error is a function of both the error probability per opportunity and the number of opportunities for an error to be made. PSFs affect the human error probability for a specific task; however, the human error frequency is also affected by the number of opportunities in a given time period. The number of opportunities for error and the human error probability are not independent. The less often a task is performed, the fewer opportunities an individual has to make a mistake. Thus, the annual error rate decreases as the task frequency decreases. However, when a task is performed less often, the individual also has fewer occasions to become skilled in the task. Thus, performing a task infrequently tends to increase the human error probability. Therefore, both factors need to be considered when choosing an IEF for a human error. The time intervals selected for Data Tables 4.3–4.5 were chosen to provide a generic IEF on a per-year basis, considering both the human error probability associated with the expected task frequency and the number of opportunities for an error to occur. For example, using Swain and Guttmann's (1983) nominal human error probability of 0.003 for an error of omission or commission, for a task containing less than 10 steps using a procedure without a checklist (Table A-1), the IEF for a human error performed daily would be 0.003 x 365 = 1.095/yr, or ~1.0/yr, as recommended in Data Table 4.3. Similarly, the IEF for a human error performed twice a month would be 0.003 x 26 = 0.078/yr, or ~0.1/yr, as recommended in Data Table 4.4.

HUMANS AS IPLs

Considerations for Human IPLs

Human factors considerations impact both the human error rate associated with an initiating event as well as the probability of failure on demand of a human IPL. When assessing a human IPL, it is wise to consider the following:

1. *Interface Factors:* Is there a suitable human-system interface such that information is presented in an easily assimilated form? Are alarms prioritized and the number of simultaneous alarms controlled to prevent alarm overload (so the human can easily perceive the critical alarm)?

2. *Workload Factors:* Does an operator's workload impact the ability of the individual to respond adequately to the alarm? Is response to an alarm prioritized above any other work that had been in progress?

3. *Training, Experience, and Familiarity Factors:* Do individuals performing the task have the training and skill level required for the activity? For high risk tasks and operator response to critical alarms, refresher training and individual performance evaluation can help ensure ongoing proficiency.

4. *Task Execution Factors:* Does a well-written procedure or troubleshooting guide exist for the alarm action that is required?

5. *Task Complexity Factors*: Do individuals performing the IPL response have the physical and mental capability to complete it successfully? Task complexity is based on the number of steps, the level of cognitive processing required, the number of people involved in the task, and the number of different interfaces with the process that are required to accomplish the task.

One essential factor in ensuring that the individual does indeed have the physical capability to complete the IPL response successfully is to validate that there is sufficient time available for the individual to diagnose the situation and complete the required response once an abnormal situation is detected. The following section discusses the key concepts related to assessing the timeline associated with a human IPL.

THE TIMELINE OF AN IPL RESPONSE

A time definition that has been used with regard to IPLs is the "process safety time" (PST). As defined in *Guidelines for Safe and Reliable Instrumented Protective Systems* (CCPS 2007b), the PST is the

"time period between a failure occurring in the process or its control system and the occurrence of the hazardous event."

As discussed in Chapter 3, there are several components of the PST. First, the process deviation becomes significant enough to be detectable by the IPL. The time required for the IPL to sense the deviation and take the required action is called the IRT (IPL response time). Finally, there may also be time required for the process to recover to a safe state; this is called the process lag time (PLT).

In the case of a human response IPL, the human has only a fraction of the PST to correctly respond to prevent the consequence of concern. It is important to select appropriate setpoints for IPLs to ensure that the operator has sufficient time to diagnose and respond to the process deviation.

A human-based IPL may or may not involve instrumentation. A safety alarm IPL has both instrumented and human components that are analyzed together to determine whether the response requirements can be met. For a human action that is not initiated by an alarm, but instead is initiated by a field reading, sample analysis, or other action demanded in a specific time frame, it is important to carefully assess the available response time. The estimation of available response time for the operator should take into account the point in the scenario timeline at which the human will likely detect the problem. The longer it takes for the operator to detect the process deviation, the less time is available to achieve a successful operator response.

Some HRA data developed for diagnosis time in control rooms (Swain and Guttmann 1983) suggest that there is a 90% chance that the diagnosis will be correct if the worker in a nuclear power plant control room has at least 10 minutes to diagnose the situation, and a 99% chance of correct diagnosis if the worker has 40 minutes. As a result of these published values, the decision time is typically set at 10 minutes. However, when giving credit for a human response, consideration should be given to potential operator reluctance to activate a safety system, when the result of activation may create a significant financial impact. History has provided examples of the consequences of human unwillingness to activate an emergency shutdown sequence that would result in significant economic impact. The Piper Alpha (Cullen 1990) and Deepwater Horizon (USCG 2011) incidents, among others, have illustrated the tragic consequences of human reluctance to activate IPLs with uncertain safety benefits but certain financial impact. It is important that the organizational culture reinforces the importance of following procedures and shutting down activities when warranted, despite potential economic consequences.

KEY POINTS

Human interaction with chemical processes is necessary for most operations. When human factors are optimized, individuals can perform reliably. However, human error has also been a major contributor to a significant number of industrial incidents. There are different types of human errors, and the error rates of various types of human error can be influenced in different ways. It is therefore important to understand the class of error that is to be prevented in order to understand how to most effectively impact the human error rate. Performance shaping factors are influences that can positively or negatively impact human performance. Data gathered from the nuclear industry and other sources have illustrated the impact that performance shaping factors have on human error rate and how good control of these factors can improve human performance.

When considering a human's role in a response IPL, consideration should be given to factors that can influence the probability of success or failure of the human response action. These factors include the human-system interface, training, workload, and experience. The individual needs to have the capability to perform the required action as well as the time necessary to successfully complete the task. The individual should be authorized to take the required action and should feel empowered to do so without waiting for supervisory approval, which can cause delays in response. Humans are often reluctant to execute response actions that they believe will have a significant financial impact on their organization. To ensure that operators don't delay in taking action due to concerns over economic consequences, it is important to have an organizational culture that emphasizes procedure use, acknowledges those who make proper decisions, and places a higher priority on process safety than on financial considerations.

REFERENCES

Baker, J., F. Bowman, G. Erwin, S. Gorton, D. Hendershot, N. Leveson, S. Priest, I. Rosenthal, P. Tebo, D. Wiegmann, and L. Wilson. 2007. "The Report of the BP US Refineries Independent Safety Review Panel; 2007." http://www. bp. com/liveas sets/bp_internet/globalbp/STAGING/global_ assets/downloads.
Bell, B., and A. Swain. 1983. *A Procedure for Conducting a Human Reliability Analysis for Nuclear Power Plants.* NUREG/CR-2254. Washington, D.C.: Nuclear Regulatory Commission.

Blackman, H., D. Gertman, and R. Boring. 2008. "Human Error Quantification Using Performance Shaping Factors in the SPAR-H Method." Paper presented at Idaho National Labs, 52nd Annual Meeting of the Human Factors and Ergonomics Society, New York, September 22-26.

CCPS. 1994. *Guidelines for Preventing Human Error in Process Safety.* New York: AIChE.

CCPS. 2007b. *Guidelines for Safe and Reliable Instrumented Protective Systems.* New York: AIChE.

CCPS. 2007c. *Human Factors Methods for Improving Performance in the Process Industries.* New York: AIChE.

Cullen, W. (Lord). 1990. *The Public Inquiry into the Piper Alpha Disaster,* Volume 1. London: HMSO.

Edwards, E., and F. Lees. 1974. *The Human Operator in Process Control.* London: Taylor & Francis Ltd.

NTSB (National Transportation Safety Board). 1990. *Marine Accident Report: The Grounding of the U.S. Tankship Exxon Valdez on Bligh Reef, Prince William Sound Near Valdez, Alaska March 24, 1989,* Washington, DC: NTSB.

Rasmussen, J., and W. Rouse. 1981. *Human Detection and Diagnosis of System Failures.* New York: Plenum Press.

Reason, J. 1990. *Human Error.* New York: Cambridge University Press.

Reason, J. and A. Hobbs. 2003. *Managing Maintenance Error: A Practical Guide.* Hampshire, U.K.: Ashgate Publishing Company.

Swain, A., and H. Guttmann. 1983. *Handbook of Human Reliability Analysis with Emphasis on Nuclear Power Plant Applications.* NUREG CR-1278. Washington, DC: U.S. Nuclear Regulatory Commission.

USCG (United States Coast Guard). 2011. *Report of Investigation into the Circumstances Surrounding the Explosion, Fire, Sinking and Loss of Eleven Crew Members Aboard the Mobile Offshore Drilling Unit DEEPWATER HORIZON in the Gulf of Mexico, April 20-22, 2010 (Volume 1).* Washington, DC: USCG.

U.S. NRC (Nuclear Regulatory Commission). 2008. *Fitness for Duty Programs - Managing Fatigue.* Regulation 10 CFR 26 Subpart I. Washington, DC: NRC.

APPENDIX B. SITE-SPECIFIC HUMAN PERFORMANCE VALIDATION

The IEF and PFD values previously presented in the data tables in Chapters 4 and 5 are considered to be reasonable values for use in LOPA, provided that human factors are sufficiently managed (refer to Appendix A). In situations where human factors are suboptimal, the values in the data tables may be optimistic; in other cases, where human factors are well optimized, these values may be conservative. Although it may indeed be possible to achieve better human performance than the values listed in the data tables would indicate, it is important to be able to substantiate a claim of lower IEF/IPL values. Regardless of whether an organization chooses to use the values in this guideline or to use alternative values for human IEFs and IPL PFDs, the organization should be able to validate the values selected.

As discussed in Chapter 2, there are several methods for developing IEF and IPL PFD values for LOPA. Values can be determined by referring to comparable data sources, by using expert judgment, and by estimating IEFs and IPL PFDs by mathematical modeling. However, it is would be preferable to collect site-specific data by directly measuring the human error rates or human IPL response failure probabilities of the workers performing the specific task being assessed.

This appendix provides examples of the data needed to develop site-specific human IEF and IPL values for use in LOPA. One key focus of this appendix is the discussion of practical means for collecting raw data in a plant setting for substantiating the error rates for the site and crediting a human IPL.

INITIATING EVENT FREQUENCY DATA COLLECTION

Before beginning an exercise in collecting human performance data, it is important to consider what types of data are needed. Although a procedure may have a number of steps, errors in performing only a few of the steps are likely to result in a LOPA scenario of concern. It would not be worthwhile to gather human error data on steps that would not result in generation of a hazard. A task analysis can be performed on the procedure to identify the critical step(s) that merit further analysis and data collection.

It is important to recognize that all human errors are not equivalent in nature. As discussed in Appendix A, some errors are errors of omission – an operator skips a step, for instance. Other errors may be errors of commission – an operator incorrectly executes an action specified in the procedure. An error in reading a value on a computer display is not the same as an error in operating a field valve. The error rates for these types of actions can differ significantly. Therefore, when gathering data, it is important to collect data on similar types of errors to make sure that the results of the analysis are valid.

Once the critical steps have been identified and the type(s) of errors to be assessed have been determined, data collection schemes can be developed for groupings of similar types of errors. The method of data collection should be more systematic than simply combining the collective memories from a group of individuals at a site. Past incident data can be useful, but it can also be quite misleading. Incidents typically occur when there is a human error *and* all IPLs fail. It would be overly optimistic to use historical incident data to represent the frequencies of human IEs. Furthermore, one rarely has any record of the number of opportunities workers had to make the error that resulted in or contributed to the incident, so it is very difficult to calculate a rate or probability. In practice, it is generally easier to show that actual site data are worse than the generic values suggested in this book, particularly if the site experiences several failures in a short period of time. It is more challenging to show that actual site data are significantly better than the values in this book; this requires a long history of data collection, with few or no failures.

Establishing tracking mechanisms for collecting data on human errors and the number of opportunities for error will yield the most reliable results. For such databases to be valid, however, a culture is needed where human error can be reported without fear of repercussion. In practice, this is very difficult to achieve.

Another means to gather human error data is by the use of simulators. This enables the site to gather statistical information on critical steps in a controlled fashion. Much of the published human error data collected in the nuclear industry were generated using simulators. While this may be a useful method for gathering human error data, its application may be limited by the availability of the technology needed to simulate the critical steps.

EXAMPLE OF SITE-SPECIFIC DATA FOR HUMAN ERROR INITIATING EVENTS

Below is an example of data generated by a site to estimate the IEF for a human error made while executing a startup procedure. The IEF is evaluated by measuring the average number of human errors of a specific type and then assessing the error rate for the particular IE of interest. The data to be collected consist of

- The number of human errors of a specific type
- The number of opportunities to make that specific type of error
- The number of steps in the procedure that could initiate a LOPA scenario
- The number of uses of the procedure per year

Based on this data, the human error probability (HEP) for a specific type of human error can be calculated:

Human error probability = (Number of observed human errors of a specific type) / (Number of opportunities to make the specific type of error)

Although a procedure can have any number of steps, a LOPA scenario is typically initiated by an error or omission in the execution of a single, specific step. The human error rate (or IEF) for a particular scenario is calculated as follows:

Human error initiating event rate (events/yr) = (Human error probability per step that could initiate the LOPA scenario) x (Number of steps in the procedure that could initiate the LOPA scenario) x (Number of uses of the procedure per year)

Example results for human error occurring during the performance of several startup procedures can be found in Table B-1. In this example, the average probability of human error per step is 0.0067 (8 errors in 1,199 opportunities) and the average error rate per initiating event is 0.0098/yr. This value would be rounded up to an IEF of 0.01/yr for use in LOPA. It would be reasonable to use this value for initiating events arising from the failure to properly complete these, and similar, procedural steps at this site.

NOTE: In some situations, no human errors may occur during a particular data collection activity. Assuming a corresponding error rate of "0" times per year in statistical calculations could be misleading and could distort the average value. In the event that no error was detected in a sample, refer to Appendix C for guidance on how to estimate a failure rate when an error has not yet occurred.

Table B-1. Example site data on human errors related to initiating event frequencies (IEFs) (based upon Bridges and Clark 2011)

Task for which a Potential Initiating Event Exists	Number of Errors of a Specific Type	Number of Opportunities	Human Error Probability (HEP)	Number of Steps of Concern	Number of Procedure Usages/yr	Error Rate per Year, per Scenario
Startup of Process A	2	184	0.0109	1	1	0.0109
Startup of Process B	3	440	0.0068	1	2	0.0136
Startup of Process C	2	220	0.0091	1	1	0.0091
Startup of Process D	1	355	0.0028	2	1	0.0056
		Average	0.0067		Average/yr	0.0098
					Site-Specific IEF for LOPA	0.01

EXAMPLE OF SITE-SPECIFIC DATA COLLECTION FOR HUMAN IPLs

Just as IEFs can be determined based on site-specific data collection, PFDs can also be determined for human IPLs for a specific site. The probability that a human will fail to act properly on demand within the available time can be evaluated by measuring the response of each responder for each action to be taken. To determine the overall PFD for the human IPL, the PFD for the human response portion of the IPL is combined with the PFD of equipment/instrumentation used for detection/annunciation of the abnormal condition and the PFD of equipment/instrumentation that the operator would use to complete the response following detection:

$$PFD_{HUMAN\ IPL} = PFD_{DETECTION\ EQUIPMENT} + PFD_{HUMAN\ RESPONSE}$$
$$+ PFD_{RESPONSE\ EQUIPMENT}$$

The sum of the PFDs for all three components comprise the overall PFD of the human IPL ($PFD_{HUMAN\ IPL}$). Although this appendix focuses on data collection to determine $PFD_{HUMAN\ RESPONSE}$, it is important to recall that the credit given in LOPA is based on the PFD of the entire human IPL.

When a limited number of operators and human IPLs are to be evaluated, it may be practical to assess each operator's performance in responding to each initiating event requiring a human IPL. However, the approach of testing every operator-human IPL combination may not be necessary to confidently establish the PFD for human response IPLs. It is also possible to determine reasonable $PFD_{HUMAN\ RESPONSE}$ values by developing a statistical plan to measure a subset of human responses for IPLs at a site. Discussed in the remainder of this appendix is a method for determining $PFD_{HUMAN\ RESPONSE}$ values by testing human responders for the actions that they are expected to take.

EXAMPLE USE OF SITE-SPECIFIC TEST/DRILL DATA TO VALIDATE HUMAN RESPONSE IPLs

Methods to collect data to validate human IPLs have not been well documented in the literature, although some chemical companies, refineries, and nuclear power plants do validate human response assumptions. Some companies also validate by direct measurement of human IPL responses based on drill data. Recent research documented the efforts to validate human IPLs using site-specific test/drill data (Bridges 2011). The following is an excerpt of the methodology employed.

Validation Setup: A simple test was used to measure the response to an alarm condition. The test was not meant to measure the probability of detection of the alarm ($PFD_{DETECTION\ EQUIPMENT}$), but rather was meant to measure the time and success in determining and accomplishing the proper response to critical alarms as part of human IPLs ($PFD_{HUMAN\ RESPONSE}$).

The test involved having multiple operators perform responses to critical process alarms that initiated human IPLs in LOPA. The accuracy of the actual response and the time to complete the response were measured to determine whether the operators were able to respond appropriately and within the allotted response time. This determined whether the operator "passed" or "failed" the test.

To run each test, the plants printed a data card (the size of an index card) and handed it to an operator chosen at random. Figure B-1 is an example of such an index card.

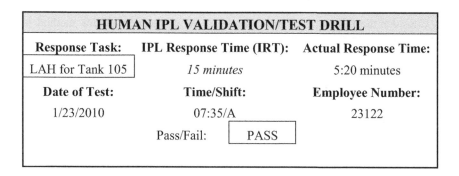

HUMAN IPL VALIDATION/TEST DRILL		
Response Task:	**IPL Response Time (IRT):**	**Actual Response Time:**
LAH for Tank 105	*15 minutes*	5:20 minutes
Date of Test	**Time/Shift:**	**Employee Number:**
1/23/2010	07:35/A	23122
	Pass/Fail: PASS	

Figure B-1. Example card used to validate a human response IPL (based upon Bridges and Clark 2011).

Note, the card contains an estimate of the IRT – the time an operator has to perform the task once the alarm is received until it is too late to take any further action. (Refer to Section 3.3.1 for a more detailed discussion of IRT.)

Validating/Testing: Generally, the tests were initiated with little warning and occurred on all shifts. The person administering the test timed the operator response and recorded the results. Each test took 10–15 minutes to administer and less than one minute to record the data. The time to respond was recorded on the index card.

A human response "failed" if the operator could not perform the required action to prevent the hypothetical outcome within the IRT. Based on the data collected, the site was able to validate that the response is accurate enough and quick enough to achieve the portion of the human IPL PFD that was allotted to human response. The testing also provided insight into the degree to which human factors were being managed on site.

APPROACH TO USING A TEST/DRILL PLAN FOR VALIDATION OF HUMAN IPLs

One method to determine whether the operators will reliably respond for each human IPL trigger is to have operators demonstrate they can individually respond to each alarm (or other trigger). This response can be demonstrated by walk-throughs in the field or by using simulators of the process unit. Some facilities may perform such drills as part of initial and refresher training, especially for procedures deemed to be critical based on the site hazard analysis. This approach has the benefit of both improving operator performance through training/drills while simultaneously gathering data to determine whether the IPL credits used in the LOPA are indeed appropriate.

Consider the following as an example of the application of this method of testing responses to IPL triggers by the operators at a chemical plant:

Background: The operating area being evaluated has 20 operators (spread across 4 rotating shifts of work). There are 130 human IPLs with similar human response requirements credited in the site LOPA.

Validation Test: The tests are documented on a set of index cards that indicate various alarm conditions; these are the events (triggers) in LOPA scenarios that should result in an appropriate response action for the given human IPL.

Demonstrated Result: A correct answer (that would indicate success of the human response portion of the IPL) is the desired response to the alarm scenario. An incorrect answer (that would indicate failure of the human portion of the IPL) is a response other than the one desired or a response that takes too long.

Estimation of Resources Requirements for This Testing Scheme: Below is an estimate of how much test effort would be needed to evaluate the human response PFD ($PFD_{HUMAN\ RESPONSE}$) for all of these identified human IPLs:

1. *Determine the number of tests to be performed.* This is calculated by multiplying the human IPLs being evaluated by the number of people who might be expected to respond as part of the human IPL at some point in the future. This example would yield 2,600 discrete tests (20 operators x 130 human IPLs = 2,600 tests) to evaluate each combination of trigger and human responder.

2. *Determine the time required to perform each test.* In this example, assuming 10 minutes of allowed response time for success per alarm/trigger, the test time for the organization would be 26,000 minutes or about 430 staff-hours (22 hours per operator per test period).

APPROACH TO USING A STATISTICAL SAMPLE PLAN FOR VALIDATION OF HUMAN IPLs

Rather than testing each operator on every human action IPL in LOPA, some organizations design sampling plans to gather statistical data. When designing such a sampling plan, it is important to ensure that human IPLs with similar triggers and operator response requirements are grouped appropriately so that the data analysis is meaningful. Refer to appropriate references (such as Walpole et al. 2006) for statistics and sampling methodologies. Note that use of this approach may require an analyst with expertise in developing statistical sampling plans.

KEY POINTS

Although a number of methods can be employed to develop IEF and human response PFD ($PFD_{HUMAN\ RESPONSE}$) data, collecting site-specific data may provide the best quality information. Near-miss databases and other systems for capturing local human error data are potential sources, although they may seriously underestimate the true IE frequency because an event typically occurs (and gets recorded) only after protection layers also fail. $PFD_{HUMAN\ RESPONSE}$ data can also be collected using drills/tests to aid in validating human IPLs.

$PFD_{HUMAN\ RESPONSE}$ can be validated by testing the response of all operators to all of the initiating events for scenarios in which they may participate as part of a human IPL. This approach, while thorough, can be time-consuming. Some organizations use a statistical sampling plan to validate the $PFD_{HUMAN\ RESPONSE}$; however, this approach will require statistical expertise to develop the plan and analyze the data. It is important that human IPLs with similar triggers and requiring similar human responses are appropriately grouped when developing a sampling plan to generate representative data that are reflective of different types of human error.

When using a test/drill protocol for evaluating $PFD_{HUMAN\ RESPONSE}$, it is important to remember that the data were collected during a simulation of a call for action. In a real event, *the stress to perform the task correctly may increase the average error rate*. It is likely not possible to conduct a drill that accurately mimics the stress of a real alarm event, so it is necessary to consider the potential impact of stress on the error rate before using data collected during drills to reduce the human IPL PFD in LOPA.

This appendix covered site-specific data collection associated with the human response portion of the IPL, $PFD_{HUMAN\ RESPONSE}$. To validate that the PFD value for the entire human IPL ($PFD_{HUMAN\ IPL}$) used in LOPA is appropriate, it is important to also consider the PFDs of any equipment/instrumentation used for the detection and annunciation of the abnormal condition ($PFD_{DETECTION\ EQUIPMENT}$), as well as the PFD of any equipment/instrumentation that the operator uses to complete the response action ($PFD_{RESPONSE\ EQUIPMENT}$).

REFERENCES

Bridges, W. and T. Clark. 2011. "LOPA and Human Reliability – Human Errors and Human IPLs (Updated)." Paper presented at 2012 AIChE Spring Meeting/8th Global Congress on Process Safety, Houston, TX, April 1-5.

Walpole, R., R. Myers, S. Myers, and K. Ye. 2006. *Probability and Statistics for Engineers and Scientists,* 8th Edition. Upper Saddle River, NJ: Pearson Publishing.

APPENDIX C. SITE-SPECIFIC EQUIPMENT VALIDATION

When using failure rate data to help estimate an IEF or IPL PFD in a LOPA, it is important to understand how equipment failure rate data are generated and the assumptions that are made so as to adapt it for use in a LOPA.

The purpose of this appendix is not to make the LOPA practitioner an expert in equipment reliability data. However, to have a meaningful discussion between the LOPA analyst and the reliability data expert, a common understanding of the general concepts and issues relating to data collection and use is necessary.

CONSIDERATIONS FOR SITE-SPECIFIC DATA COLLECTION

Obtaining high-quality equipment reliability data has traditionally been a difficult task due to the lack of adequate tools for the initial data capture, the required manipulation/analysis of the data, and the need for integrating the input of multiple functional areas of expertise.. To compound the problem, gathering high-quality data is generally not the highest priority for any of the groups involved. With an emphasis on data quality, and with the availability of computerized databases, many of the traditional barriers (such as personnel time requirements for data capture and data analysis) are diminished. This greatly improves the business case for data collection. Below is a systematic approach that allows the various functional disciplines to work jointly to develop high-quality data. For greater insight, refer to the ongoing CCPS Process Equipment Reliability Database (PERD) initiative (CCPS n.d.).

The general steps necessary to generate high quality data are as follows:

1. Identify the need to generate site-specific data, communicate intent and benefit to management, and gain support.

2. Identify the equipment of interest.

3. Identify SMEs (Subject Matter Expert) for the equipment of interest and allocate sufficient SME resources.

4. Form a technical guidance team with the SMEs identified.

5. Develop an equipment taxonomy (reference the ongoing CCPS PERD initiative).

6. Identify and document potential or experienced failure modes as per the taxonomy.

7. Identify the pertinent data for analysis.

8. Perform a gap analysis compared to existing data and revise (if possible) existing methodology/database(s) to provide for data capture.

9. Establish an efficient work process for data collection and analysis.

10. Develop and document a data quality collection plan and train mechanics/technicians on the pertinent details of data collection.

11. Leverage applicable IT management systems, maintenance management system, etc.

12. Establish an offline database for the transfer of data identified as necessary for analysis. Reference technical information documented by the ongoing CCPS PERD initiative.

13. Collect the appropriate data with the work process documented in step 9 to ensure the defined level of quality assurance.

14. Analyze the data to determine applicable reliability parameters and key performance indicators, such as equipment failure rate, mean time in bypass, mean time to repair, mean time to restore, etc.

To better understand this process, it is useful to understand the concept of taxonomy. Equipment taxonomy is a method of classification where components are organized in similar groups and subsets of these components are listed beneath their parent components. Equipment taxonomy is described as a hierarchical organization of data cells, where the items contained in a given level have more equipment reliability characteristics in common with each other than they do with items in any other level (CCPS 1989). In other words, taxonomy provides a logical method for the classification of various equipment populations. A taxonomy definition is required when using this method to compile data tables (such as used in *Guidelines for Process Equipment Reliability Data – with Data Tables* (CCPS 1989), *Offshore Reliability Data Handbook,* 5th Edition, Volume 1 – *Topside Equipment,* Volume 2 – *Subsea Equipment,* 5th Edition (OREDA 2009), and *Guide to the Collection and Presentation of Electrical, Electronic, Sensing Component, and Mechanical Equipment Reliability Data for Nuclear-Power Generation* Stations (IEEE 1984).

Another consideration when estimating equipment performance is determining the useful life of the equipment. For equipment, the term "useful life" is a statistical measure that represents the length of time in which the failure rate is expected to be relatively constant. This time ends where the failure rate is no longer constant, but is instead accelerating due to wear and age. *Useful life is not to be interpreted as the time at which equipment must be replaced.* While replacement may be appropriate for some equipment, other equipment can be repaired or refurbished. Rebuilding compressors and relief valves are examples of maintenance activities that also extend the useful life of the equipment. Knowledge of the useful life of the components comprising the equipment is important in developing the ITPM program, since proof test intervals should not extend beyond the statistical useful life of any component.

In the process industries, typical statistical useful life values for equipment that are routinely used for protection can range from less than a year to decades (compared with some plant life expectations in excess of three decades). It is important to ensure that the frequency of testing is appropriate for the expected useful life of the equipment in the operating environment. This may require more frequent testing than the off-line maintenance frequency for the process equipment. This mismatch may be resolved through implementation of various measures, including provision for on-line testing and maintenance. ISA TR84.00.03 (2012) provides guidance on managing an ITPM program for SIS; this guidance is generally applicable to other safety controls, alarms, and interlocks (SCAI) as well.

For random failure rate data to be valid throughout the life of the facility, it is assumed in the mathematical calculations that the equipment is sufficiently maintained so that it continues to operate within its "useful life" condition and will not reach wear-out prior to its next inspection/proof test. This is often referred to as returning the equipment to a "like new" or "as good as new" condition. Effective proof tests involve more than exercising the protection layer to see if it functions. The equipment should also be inspected for incipient conditions. These are conditions that, if left unrepaired, could likely lead to failure prior to the next proof test. Checking the conditions of related devices is also important to the overall reliability of the system. The basic premise for using the random failure rates includes effective identification of all dangerous failure modes, designing the proof test and inspection program with the ability to detect those failures, and correcting them before putting the equipment back into service. Any deviation from this premise will adversely affect the ability of the IPL to be effective.

In order for inspection and testing data to be useful, it is important to provide guidance regarding the pass/fail criteria. It also requires that the person developing the inspections and tests understands what incipient conditions and failure modes are of significance. Too often, it is assumed that the mechanic or technician understands when action needs to be taken in response to an as-found condition. Unfortunately, without the guidance of well-written maintenance procedures and training in the execution of these procedures, this assumption generally leads to inconsistent performance.

As part of defining pass/fail criteria, the applicable maintenance and inspection procedures are documented and specific actions are expected for factual conditions found. Even when there are documented procedures, the quality of these procedures can be suspect if proper consideration is not given to failure modes important to the application. This affects the test efficacy and the quality of the data reported. Consider the following example (Arner and Thomas 2012).

Example: Suppose there is a ball-type emergency shutoff valve that is required to close in the event of a high-pressure demand. In this example, assume that seat leakage is not a concern; the important consideration is that the valve closes. Often, in this situation, the pass/fail criteria are to visually confirm movement of the valve stem. However, although the stem sufficiently moves, there is no confirmation that the ball moves (or is even present)

Although selecting a ball assembly based on prior use in the operating environment reduces the likelihood of failure, there is still a failure rate for the valve not closing. One means to confirm valve closure is to perform a valve leak test, even when tight shutoff is not required by the application. In addition, valves are often tested (stroked) when the plant is shut down.

Unless pressure exists upstream of the valve that would be normally present during a real process demand, it cannot be demonstrated by a simple stoke test that the actuator will have sufficient force available to close the valve. As before, the failure mode can be reduced but not eliminated. Providing the means to pressurize the valve inlet during the aforementioned leak test via pressure pumping the volume upstream of the valve is an effective way to ensure that a valve continues to maintain its capability to provide closure under actual conditions.

ESTIMATING FAILURE RATES AND PROBABILITIES USING GENERIC DATA

At the end of this appendix are references to a variety of generic data. These references may contain predicted data, generic data based upon some level of operating experience, data based upon expert judgment, data specific to selected sites, data from vendors, or data from any combination of these sources.

The data tables in this document are composed primarily of generic data. The entries include IEs or IPL PFDs for which the *Guidelines* subcommittee: (1) had data, and (2) developed consensus. Other IE frequencies and IPL PFDs may not be provided in this document for any number of reasons, as discussed in Chapters 4, 5, and 6. When data are not shown in this document for a particular application, it may be necessary to use other sources of generic data to generate the needed values. Guidance is provided in Chapter 4 on IE and in Chapter 5 on IPL regarding how to develop values beyond those already presented.

The following example illustrates a method for deriving and converting generic data for use as IEF and/or IPL PFD.

Example 1: Estimating the average PFD from available failure rate data. PFD_{avg} represents the average PFD over the proof test interval of an equipment item. Table C-1 illustrates the estimation of the total PFD for a high pressure shutdown system, given that the failure rates of each individual component and the test frequencies of each component are known.

Table C-1. Estimation of PFD_{avg} of a high pressure shutdown system

Component	Dangerous Failure Rate, λ	Reference	Proof Test Interval	**PFD_{avg}
Ball valve Actuator: pneumatic diaphragm, spring return	2.80×10^{-2} / year	*Safety Equipment Reliability Handbook*, 3rd Edition, exida, 2007	3 years	4.20×10^{-2}
3-way solenoid valve fails to vent when de-energized	1.67×10^{-2} / year	SIL Solver®, 2012, Low wattage solenoid, De-energize to trip service	3 years	2.50×10^{-2}
Trip Amplifier	3.5×10^{-4} / year	PDS Data Handbook (SINTEF 2010), p 86.	3 years	5.25×10^{-4}
Direct process connected pressure transmitter	2.63×10^{-3} / year	*OREDA-2009: Offshore Reliability Data*, p 467 (OREDA 2009)	3 years	3.94×10^{-3}
			PFD Total	7.15×10^{-2}

**PFD Contribution calculated using the simplified equation $PFD_{avg} = (\lambda\ T)/2$

NOTE 1: Use of the simplified equation requires that the product of λT is less than or equal to 0.1, which can be achieved through appropriate design and management.

NOTE 2: Use of the simplified equation assumes maintenance effectiveness and probability of human error have been properly managed to ensure the equipment is returned to service in "useful life condition sufficient to not reach wear out prior to the next inspection/proof test. Readers wishing to gain a more in depth understanding of these issues is referred to ISA TR84.00.02, *Safety Integrity Level (SIL) Verification of Safety Instrumented Functions* (ISA 2002).

ESTIMATING FAILURE RATES AND PROBABILITIES USING PREDICTED DATA

Predicted data for total failures of equipment are developed using techniques such as the parts count or parts stress methods in *Military Handbook 217F* (U.S. DoD 1990). Failure rate data are more likely to be available for components, such as transistors, capacitors, springs, and bearings, than for the equipment as a whole. By summing the failure rates of the individual components, it is possible to predict an overall failure rate for the equipment. Failure Mode and Effects Analysis (FMEA) can be used to identify the failure modes of each component and determine the effect of each failure mode on the equipment or process. The failure rate associated with each failure mode can then be determined using predicted data. The predicted failure rate values, whether for the equipment as a whole or for the individual failure modes, are highly dependent upon the assumptions made in the analysis. Failure rates can vary considerably as a function of the actual operating environment as compared to the environment in which the generic component data were collected and tabulated.

Predicting failure rates using this method is complex and requires expertise in the detailed aspects of the equipment design and failure modes. As an example, IEC 61508 (2010) addresses the evaluation of programmable electronic safety equipment used in SIS applications and includes specific requirements for the use of predictive techniques. Consequently, many manufacturers make failure rate claims for their products based on these techniques. Predictive failure rate values are useful for comparison with actual plant experience values to ensure that products are capable of providing the expected performance, which should reflect the operating environment characteristics. Although the user may not be a specialist, it is important to understand the basics presented in this section. To be an informed consumer, it is necessary to understand the decisions of manufacturers with respect to the assumptions they make in their predictions and to understand the limits of their equipment boundary.

ESTIMATING COMPANY-SPECIFIC FAILURE RATES AND PROBABILITIES USING PLANT-SPECIFIC DATA

The methods of estimation and evaluation employed depend upon whether the equipment failures are revealed or unrevealed, as described in Section 3.4.3. For equipment with revealed failures, it is important to understand whether the equipment is repairable and whether the data quality is sufficient for LOPA purposes. Quality of the data is inversely related to the number of assumptions that must be made about the data. In this context, quality is related to a number of factors:

- How general or specific are the groupings of devices? Do the data groupings differentiate between different types and sizes of pumps, or are all pumps of all sizes grouped together?

- What is the size of the grouping? The larger the grouping, the more likely it is that the data will be statistically meaningful.

- Do the data groupings differentiate between key attributes, such as time in service and process severity? These attributes can significantly impact the failure rate for a given component.

- What was the environment in which the component functioned? Environmental conditions can vary from mild to corrosive, and differentiation between data in different operating environments typically yields data of higher quality.

In general, the more specific the data groupings, the easier it is to obtain higher quality data that are more pertinent to a specific component in a specific process service and operating environment. For a more detailed discussion on estimating failure rates and probabilities of failure, the reader is referred to Chapters 3 and 4 of *Guidelines for Improving Plant Reliability through Data Collection and Analysis* (CCPS 1998a).

ESTIMATING FAILURE RATE WHEN THE FAILURE HAS NOT YET OCCURRED

There are times when limited data are available because no failures or events have yet occurred. In these situations, the analyst should determine whether the event of concern is physically possible. If the event cannot occur due to the laws of physics, the event should not be considered in the LOPA study. Where the event is physically possible and no events have occurred in a time period of n years, the equation below provides a conservative estimate of the failure rate:

$$\lambda_{event} = 1/n$$

where

λ_{event} = Estimated failure rate of the event of concern, events per year

n = Number of years without occurrence of the failure or event

Example calculation: If there have been no observed or detected failures in 11 years, then an estimate of the λ_{event} = $1/11$ yr = $0.09/yr$. This is equivalent to assuming one failure in the period.

There are two caveats regarding this estimate:

- It is recommended that a minimum of 10 years of data be available, during which time the event of concern has not happened, before this estimate is applied.

- Systems where the failure will not be revealed until a test is completed should be examined carefully. The analyst should verify that the test protocol will find the failure of concern. If the test protocol is not capable of determining the status of the system, this implies that the failure could have occurred and not been detected during the time period. In this case, no credit should be assigned for the related IPL.

The analyst may refine the estimate using the rule of $1/(3n)$ (Welker and Lipow 1974). The frequency calculated using this method attempts to estimate the mean failure rate, as opposed to the upper bound in the $1/n$ method.

$\lambda_{event} = 1/(3n)$

Example calculation: If there have been no observed or detected failures in 11 years, then the estimate of $\lambda_{event} = 1/(3 \times 11\ yr) = 0.03\ yr$.

More sophisticated methods are available and can be used in more advanced studies, such as in a quantitative risk analysis (Bailey 1997; Freeman 2011).

SELECTED EXAMPLE FOR CALCULATING RELIABILITY DATA FOR USE IN LOPA FROM PLANT-SPECIFIC DATA

In one company, the operating group was able to provide the following data on self-contained temperature control valves used as an IPL:

- Conservatively estimated installed population = 3,490

- The tracking system detected two fail-closed valves per month over a period of three years, or 24 failures per year. There was confidence in these numbers as the failures were revealed, causing loss of supply to the customer.

- The estimated fail-closed rate, $\lambda_{fail\ closed} = 24/3490 = 0.0069$ fail-closed valves per year.

The failure mode of interest was failing open; unfortunately, it is an unrevealed failure and data did not exist. From FMEA, it was known that the equipment tended to fail closed rather than open. With this information, it was conservative to assume the same failure rate for the valve to fail open as for the valve to fail closed, i.e., 0.0069/yr. To convert that value into a PFD_{avg}, assuming a 5-year proof test interval, the following simplified calculation was performed:

$PFD_{avg} = (\lambda T)/2 = (0.0069)(5)/2 = 0.017$

where

λ = Failure rate

T = Proof test interval

SOURCES OF DATA

The following list provides additional sources of generic data and/or more detailed explanation of some of the more advanced topics relevant to equipment failure rate data, failure modes, data collection to support "proven-in-use," and the methods to analyze such data:

1. *Process Equipment Reliability Database (PERD).* (CCPS n.d.). Readers wanting to know more about developing high-quality, reliable data and measuring performance on a cost-effective basis are referred to the website describing the CCPS ongoing PERD initiative: http://www.aiche.org/CCPS/ActiveProjects/PERD/index.aspx

2. *Guidelines for Process Equipment Reliability Data – with Data Tables* (CCPS 1989). The book provides descriptions of a number of data sources covering both chemical and nonchemical industry databases, sources, and studies. In addition, it contains useful technical background information regarding equipment taxonomies as well as a section covering data collection to support "proven-in-use" analysis.

3. *Guidelines for Improving Plant Reliability through Data Collection and Analysis* (CCPS 1998a). This book provides a foundation for understanding failure modes and the quality assurance work process needed to support quality data for proven in use analysis. In addition, this book provides a more in-depth discussion of failure data analysis.

4. OREDA *Offshore Reliability Data Handbook,* 5th Edition, Volume 1 – *Topside Equipment,* Volume 2 – *Subsea Equipment,* 5th Edition (OREDA 2009). This reference contains data for 2000–2003 for a broad range of process equipment used on offshore oil drilling platforms. The data are "proven-in-use" for the specific application viewpoint; however, they are generic in the sense that it is not possible to determine any differences that may exist due to the quality of asset integrity programs employed by the different companies contributing the data. Each edition [1st, 2nd, 3rd, and 4th] covers different years of data from 1984 to 2002.

5. IEEE Standard-493-2007, *Recommended Practice for the Design of Reliable Industrial and Commercial Power Systems* (2007). Known as the "Gold Book," it contains a variety of electrical equipment failure rate data that were determined via surveys of industrial and commercial installations.

6. IEEE Standard-500, *Guide to the Collection and Presentation of Electrical, Electronic, Sensing Component, and Mechanical Equipment Reliability Data for Nuclear-Power Generation Stations, reaffirmed 1991* (1984). A method of collecting and presenting reliability data for use in nuclear power generating station reliability calculations is established. It applies to reliability data for electric, electronic, sensing component, and

mechanical equipment. (This standard has been withdrawn and is no longer updated.)

7. *Safety Equipment Reliability Handbook*, (exida, 2007). Data in this handbook are largely the result of predictive analysis by quantification of Failure Modes, Effects, and Diagnostic Analysis studies. Equipment addressed is typically intended for use in SIS in accordance with IEC 61511 (2003).

8. *SIL Solver®* (SIS-TECH 2012). Data in this database are the result of a Delphi process that uses expert judgment to select appropriate failure rate values for typical process operating environments. Equipment addressed is typically intended for use in control systems and SCAI, and SIS. The database provides estimates of dangerous and spurious failure rates to support the calculation of the probability of failure on demand and spurious trip rate for instrumented systems.

9. *Electrical & Mechanical Component Handbook* (exida 2006). Data in this handbook are largely the result of predictive analysis using IEC 62380 (2004). This is one of the few data sources that provides estimated useful life values in addition to failure rates.

10. ISA TR84.00.02, *Safety Instrumented Functions (SIF) – Safety Integrity Level (SIL) Evaluation Techniques* (2002). Data in this database were collected from ISA committee members. Equipment addressed is intended for use in SIS. The database provides estimates of dangerous and spurious failure rates for a limited number of components.

11. *PDS Data Handbook* (SINTEF 2010). This handbook contains reliability data for input devices, logic controllers, and final elements for use in SIS design and verification.

12. *EPRD – Electronic Parts Reliability Data* (RIAC 1997). This document contains reliability data on both commercial and military electronic components for use in reliability analyses. It contains failure rate data on integrated circuits, discrete semiconductors (diodes, transistors, optoelectronic devices), resistors, capacitors, and inductors/transformers, all of which were obtained from the field usage of electronic components.

13. *SPIDR™ – System and Part Integrated Data Resource* (SRC 2006). This is a database of reliability and test data for systems and components.

14. *SR-332 – Reliability Prediction for Electronic Equipment* (Telcordia Technologies 2011). This reference provides tools needed for predicting device and unit hardware reliability.

15. *FARADIP – Electronic, electrical, mechanical, pneumatic equipment* (M2K 2000). This source provides failure rate and failure mode data banks based on more than 40 published data sources, together with M2K's own reliability data collection. It provides failure rate data ranges for a nested hierarchy of items covering electrical, electronic, mechanical, pneumatic, instrumentation and protective devices. Failure mode percentages are also provided.

16. ISA TR84.00.03, *Mechanical Integrity of Safety Instrumented Systems (SIS)* (2012). The technical report provides considerations for establishing a mechanical integrity program for safety instrumented systems and focuses on how to plan and implement a comprehensive program.
17. Military Handbook 217F – *Reliability Prediction of Electronic Equipment* (U.S. DoD 1990). The handbook establishes and maintains consistent and uniform methods to estimate the inherent reliability electronic equipment and systems. It provides a common basis for reliability predictions of this equipment. The handbook contains two methods of reliability prediction, the Parts Count and Part Stress methods (where the Parts Count method is generally more conservative). As part of the handbook, several failure rates are available for parts such as integrated circuits, transistors, diodes, resistors, capacitors, relays, switches, connectors, etc.
18. FMD-91 – *Failure Mode/Mechanism Distributions* (RIAC 1991). This document presents failure mode distributions intended to support reliability analysis such as Failure Mode and Effects Analysis (FMEA) and Failure Modes, Effects, and Criticality Analysis (FMECA). It is useful in conjunction with other data sources that simply list a total failure rate, such as the Military Handbook 217F (U.S. DoD 1990) and NPRD-95 (RIAC 1995).
19. NPRD-95 – *Nonelectronic Parts Reliability Data* (RIAC 1995). This document contains failure rates for a variety of electrical, electro-mechanical and mechanical parts and assemblies.

REFERENCES

Arner, D., and H. Thomas. 2012. "Proven in Use (What's the Quality of Your Data?)." Paper presented at 8th Global Congress on Process Safety, Houston, TX, April 1-4.
Bailey, R. 1997. "Estimation from Zero Failure Data." *Risk Analysis* Vol. 17. No. 3:375-380.
CCPS. 1989. *Guidelines for Process Equipment Reliability Data, with Data Tables.* New York: AIChE.
CCPS. 1998a. *Guidelines for Improving Plant Reliability through Data Collection and Analysis.* New York: AIChE.
CCPS. n.d. Process Equipment Reliability Database (PERD). http://www.aiche.org/ccps/resources/perd.
Exida. 2006. *Electrical & Mechanical Component Handbook.* Sellersville, PA: Exida.
Exida. 2007. *Safety Equipment Reliability Handbook,* 3rd Edition. Sellersville, PA: Exida.
Freeman, R. 2011. "What to Do When Nothing has Happened," *Process Safety Progress,* September, Vol. 30, No. 3: 204-211.
IEC (International Electrotechnical Commission). 2003. *Functional Safety:*

Safety Instrumented Systems for the Process Industry Sector - Part 1: Framework, Definitions, System, Hardware and Software Requirements. IEC 61511. Geneva: IEC.

IEC. 2004. *Reliability data handbook – Universal model for reliability prediction of electronics components, PCBs and equipment.* IEC 62380. Geneva: IEC.

IEC. 2010. *Functional safety of electrical/electronic/programmable electronic safety related systems.* IEC 61508. Geneva: IEC.

IEEE (Institute of Electrical and Electronics Engineers). 1984. *Guide to the Collection and Presentation of Electrical, Electronic, Sensing Component, and Mechanical Equipment Reliability Data for Nuclear-Power Generation Stations.* Standard 500-1984 reaffirmed 1991. New York: IEEE.

IEEE. 2007. *Recommended Practice for the Design of Reliable Industrial and Commercial Power Systems.* Standard 493-2007. New York: IEEE.

ISA (International Society of Automation). 2002. *Safety Instrumented Functions (SIF) – Safety Integrity Level (SIL) Evaluation Techniques.* TR84.00.02-2002. Research Triangle Park, NC: ISA.

ISA. 2012. *Mechanical Integrity of Safety Instrumented Systems (SIS).* TR84.00.03-2012. Research Triangle Park, NC: ISA.

M2K (Maintenance 2000). 2000. *FARADIP – Electronic, electrical, mechanical, pneumatic equipment.* [computer program] http://www.m2k.com/failure-rate-data-in-perspective.htm.

OREDA. 2009. *Offshore Reliability Data Handbook,* 5th Edition *(OREDA).* Trondheim: SINTEF.

RIAC (Reliability Information Analysis Center). 1991. *Failure Mode/Mechanism Distributions.* FMD-91. Rome, NY: RAC.

RIAC. 1995. *Nonelectronic Parts Reliability Data.* NPRD-95. Rome, NY: RAC.

RIAC. 1997. *Electronic Parts Reliability Data (EPRD).* NPRD-97. Rome, NY: RAC.

SINTEF (Stiftelsen for industriell og teknisk forskning). 2010. *PDS Data Handbook.* Trondheim, Norway: SINTEF.

SIS-TECH. 2012. *SIL Solver®.* (Version 6.0) [computer program] http://sis-tech.com/software.

SRC (Alion System Reliability Center). 2006. *SPIDR™ – System and Part Integrated Data Resource* [computer program] http://src.alionscience.com/spidr/.

Telcordia Technologies. 2011. *Reliability Prediction for Electronic Equipment.* SR-332. Morristown, NJ: Telcordia.

U.S. DoD (Department of Defense). 1990. *Reliability Prediction of Electronic Equipment.* Military Handbook 217F. Washington, DC: U.S. Department of Defense.

Welker, E., and M. Lipow. 1974. *Estimating the Exponential Failure Rate from Data with No Failure Events.* Proceedings of the 1974 Annual Reliability and Maintainability Symposium, Los Angeles, California. New York: IEEE.

APPENDIX D. EXAMPLE RELIABILITY DATA CONVERSION FOR CHECK VALVES

This appendix provides a rough estimate of the probability of failure for a check valve in both low demand usages (demands of less than 1/yr) where a check valve may serve as an IPL, and for high demand services where a check valve failure may serve as an initiating event for a large backflow. The component data were used to look at 1oo1 (one out of one, or a single check valve) and 1oo2 (one out of two redundant check valves in series) architectures. For each of these architectures, results were estimated for proof test intervals ranging from one to six years.

DATA DISCUSSION

The mean probability of failing to close is listed in the source material *Reliability Data Book for Components in Swedish Nuclear Power Plants* (Bento et al. 1987), as 0.0034 per demand and the upper limit at the 95% confidence level was listed as 0.0019 per demand, where the proof test interval was one year. A review of the data indicates that the value is predominantly based on the failure of the check valve to prevent large backflow, rather than small leaks, and is listed as failure to close.

DATA CONVERSION TO FAILURE RATE

The following analysis assumes that the exponential distribution is applicable and uses the following equation for calculating the probability of failure at the end of the time interval, based on a constant failure rate λ and a proof test interval time t:

$$P = 1 - e^{-\lambda t}$$

$$e^{-\lambda t} = 1 - P$$

$$Ln\ (e^{-\lambda t}) = Ln(1 - P)$$

$$-\lambda t = Ln(1-P)$$

$$\lambda = Ln(1 - P)/(-t)$$

where

P = Probability of failure (on demand); equivalent to PFD_{avg}

t = Time; in this instance we will substitute T, the test interval (time between tests), expressed in years

λ = Failure rate; the number of failures expected or experienced per year

Inserting the available data yields the following:

$\lambda = Ln(1 - 0.019)/(-1\ yr)$

$\lambda = 0.019$ failures/yr

FAULT TREE ANALYSIS SUMMARY RESULTS

The component data derived above were used to look at 1oo1 (one out of one, or single check valve) and 1oo2 (one out of two, or redundant check valves in series) architectures. For each of these architectures, results were obtained for proof test intervals ranging from one to six years. Summaries of the results are included in Tables D-1 and D-2.

NOTE: The same installation and maintenance criteria shown in Chapters 4 and 5 for single and double checks are also required to use this method of conversion of data, unless the site has installation and maintenance data and criteria that, while different, yield acceptable reliability.

The failure rates and PFDs for the two cases, of a single check valve and dual check valve, are shown:

Table D-1. Single check valve – failure rates and derived PFD

Architecture: 1oo1 (one out of one)

Failure Mode: Fail to close

Proof Test Interval, Years	Failure Rate per Year	PFD (Average)
1	1.92E-2	9.6E-3
2	1.92E-2	1.9E-2
3	1.92E-2	2.8E-2
4	1.92E-2	3.8E-2
5	1.92E-2	4.7E-2
6	1.92E-2	5.6E-2

Table D-2. Two check valves in series – failure rates and derived PFD

Architecture: 1oo2 (one out of two)

Failure Mode: Fail to close

Common Cause Factor: Assumed to be zero for this example, *but common-cause failures should be considered in the event that they dominate the resultant PFD in the specific application.*

Proof Test Interval, Years	Failure Rate per Year	PFD (Average)
1	3.7E-4	1.2E-4
2	7.1E-4	4.8E-4
3	1.0E-3	1.1E-3
4	1.4E-3	1.9E-3
5	1.7E-3	2.9E-3
6	2.0E-3	4.1E-3

GUIDANCE FOR LOPA AND QRA

When check valves are being used in protective service, it is important to determine whether they are in low demand or continuous/high demand service. For a more complete explanation of low and continuous/high demand service, refer to Chapter 3 Sections 3.5.1 and 3.5.2 of this document or Annex I of ISA Technical Report TR84.00.04, *Guideline on the Implementation of ANSI/ISA 84.00.01-2004*, Part 1 (ISA 2011).

For continuous demand cases, the check valve failure rate should be used as the initiating event rate. The check valve can be considered to be in continuous demand service if

- The check valve is the only protection available to prevent the consequence of concern, or
- The check valve failure is the initiating cause that allows backflow to result, and the demand rate is greater than once per year.

If the failure rate is not known directly, it can be estimated based on a known PFD value and test interval, as demonstrated in the "Data Conversion to Failure Rate" section earlier in this appendix.

For low demand service, the check valve probability of failure should be used as the PFD for the backflow prevention IPL. The check valve can be considered to be in low demand service if the demand rate on the check valve is less than once per year.

REFERENCES

Bento J.-P., S. Björe, G. Ericsson, A. Hasler, C.-D. Lydén, L. Wallin, K. Pörn, O. Akerlund. 1987. *Reliability Data Book for Components in Swedish Nuclear Power Plants*. RKS/SKI 85-25. Stockholm: Swedish Nuclear Power Inspectorate and Nuclear Training & Safety Center of the Swedish Utilities.

ISA (International Society of Automation). 2011. *Guidelines for the Implementation of ANSI/ISA 84.00.01*. TR84.00.04-2011 Part 1. Research Triangle Park, NC: ISA.

APPENDIX E. PRESSURE VESSELS AND PIPING OVERPRESSURE CONSIDERATIONS

Estimating the severity of overpressure is a common activity in risk analysis; and yet it seems to pose considerable problems for many risk analysts and is therefore deserving of special emphasis.

DEFINING OVERPRESSURE

As the pressure within a system rises, the probability of a leak or a rupture increases, although the specific failure point is dependent on system design and maintenance. There are, however, some issues to consider when assigning the consequence category to an overpressure event. Vessels and piping design can provide protection against overpressure when appropriate design stress calculations and construction codes are used.

For example, the ASME Boiler and Pressure Vessel Code (BPVC) Section VIII (2013) uses the concept of Maximum Allowable Working Pressure (MAWP). This pressure is determined based on the design temperature, allowable stress for each material in the system, and the installed thickness minus the corrosion allowance. The code dictates the inclusion of a safety factor in the calculation of the maximum pressure that the vessel can continuously withstand. Some pressure vessel codes use a design margin of 3–4 times the material strength required at the design pressure (ASME 1999, ASME 2013). The vessel is also typically hydrotested at a pressure of 1.3–1.5 times the MAWP. Other boiler and pressure vessel codes will use similar concepts but may have different design margins.

The 2013 ASME code allows the relief valves protecting the vessel to have a margin above the MAWP, to account for the characteristics of the relief valve as it opens and to allow operational flexibility. Thus, using the 2013 ASME code as an example, a single relief valve is required to be capable of limiting the maximum pressure to no more than 110% of MAWP. Multiple relief valves are required to be capable of limiting the maximum pressure to no more than 116% of MAWP. For a single or multiple relief valves in a fire scenario, the size is required to be large enough to limit the maximum pressure to no more than 121% of MAWP. Also, the codes require that the vessel be pressure tested at greater than the MAWP.

A sudden pressure rise, such as one due to a runaway reaction or internal deflagration, can cause a large hazardous material release and/or other severe consequences due to overpressure rupture of a vessel. Conversely, while it is certainly true that gradually increasing internal pressure increases the probability of a leak from flanges, packing, seals, and instrument connections, it is also true that normally these relatively small leaks may also help to prevent catastrophic rupture of the system by providing venting of excessive pressure. Such a release will typically have local, rather than widespread, effects. Consequences that result from the release, such as local fire, small vapor cloud, or liquid pool will also tend to be localized and have limited effects. However, release of highly toxic materials can still lead to localized fatalities, even for small leaks. The combination of design criteria and implementation of properly designed relief systems has proven effective such that overpressure ruptures of a vessel and piping systems are rare.

Some companies may define their policy such that any increase in pressure above the allowable overpressure is a violation of code and could choose to assume catastrophic failure when it is exceeded. This is a conservative approach, and it results in scenarios in which only small increases in pressure beyond the MAWP will be treated as equivalent to scenarios where there is a significantly higher potential for a rupture.

FACTORS THAT LIMIT PRESSURE RISE

Initiating events and the failure of IPLs may result in scenarios that lead to an energy imbalance in a system, with a corresponding pressure rise. However, this does not necessarily mean that the pressure rise will be uncontrolled or unlimited. The characteristics of the equipment, and the vapor-liquid equilibrium of the material in the process, often act to limit the maximum pressure that is reached in the system. Consideration should be given to these issues to support the realistic estimate of consequence severity assigned to a scenario.

Examples of factors that can affect the maximum pressure rise in equipment are listed below. Note that the factors below only apply to properly maintained and inspected equipment that have not degraded below the thickness/integrity required for the design intent.

  • *Upstream pressure* – If the upstream pressure is limited, then the maximum pressure that the system can experience is also limited. For example, if the IE is a failure of a pressure control system, and the supply pressure could be no more than 300 psig through inherent means, then the maximum pressure that the system could reach would be 300 psig. If the MAWP for the system was 230 psig, then the overpressure would be approximately 30%. This would be above any code pressure limit (such as MAWP), but below a typical hydrotest pressure of 130–150% of the MAWP. Using the most conservative approach, this could be treated as a rupture. However, a reasonable approach could also be to state the consequences as "exceeding of code pressure limit and possibility of leak from flange or instrument connection." (Some organizations might not even consider this a LOPA scenario, since the expected overpressure is still below the test pressure.)

- *Centrifugal pump maximum discharge pressure* – This is a particular case of limited upstream pressure. For example, suppose that a new impeller has been installed for a pump and the shutoff head is now 130 psig, as compared to the original 100 psig. The system connected to the pump had a MAWP of 100 psig. If the pump is deadheaded due to a valve closing downstream of the vessel being fed by the pump, then the pressure would rise to 30% above MAWP. The probable consequence would not be a rupture, but instead would most likely be a leak from a flange or a seal or an instrument connection. This approach is *not* meant to suggest ignoring code requirements; however, a range of consequences "begin" when exceeding the code requirements, and the potential consequences will gradually escalate as the pressure continues to rise.

  For a positive displacement pump, the analyst would need to be careful to use the true maximum discharge pressure. Positive displacement pumps will generate discharge pressures limited only by the capacity of the system to contain the pressure or tolerance in the pump itself. Also, analysts should be aware that a centrifugal pump can generate pressure greater than its maximum discharge pressure if it is isolated. The energy input from the spinning impeller can become a source of heat, causing thermal expansion of the liquid. The additional heat can also initiate a thermal decomposition or other uncontrolled reaction in some systems.

- *Compressor* – A similar situation to the centrifugal pump can arise for centrifugal compressors.

- *Temperature pinch for distillation column bottoms reboiler* – When a distillation column experiences a pressure rise, the boiling point of the column bottoms also rises due to the vapor-liquid equilibrium of the system. As the boiling point rises, such as from a change in composition or increased pressure, it narrows the temperature differential between the heating media (usually steam) and the column bottoms in the reboiler. This causes the rate of energy input into the column to decrease. If the pressure continues to rise, the boiling point of the column bottoms will eventually reach the temperature of the heating media for the reboiler. Unless there is additional energy input to the column from another source, the pressure at which the heating media temperature for the reboiler and the column bottoms temperature are equal represents the maximum pressure that can be reached, assuming a nonreactive system.

  As an example, suppose that, for a given column, the MAWP is 300 psig. In the event of loss of temperature control, the steam to the reboiler could potentially heat the column bottoms to the boiling point, resulting in 375 psig pressure in the column. In this case, the maximum pressure that could be achieved if the relief valves did not open, and assuming no other energy input, would be approximately 125% of the MAWP. This may be a code violation but is not likely to result in a rupture of the system, unless the integrity of the system is already severely compromised, such as by corrosion or embrittlement. A leak from a flange, seal, or instrument connection would be more likely; although, no vessel rupture would likely occur.

OPTIONS FOR TREATMENT OF OVERPRESSURE

There is no definitive approach throughout the process industries for treatment of overpressure within the context of risk assessment. Rather, individual companies have developed their own approaches. To help companies determine which approach would be most suitable, various options are presented below, along with the associated advantages and disadvantages:

Option 1: Assume any pressure rise that exceeds code criteria will result in a rupture with major direct and indirect consequences in terms of human fatalities and injuries, capital and operating losses, and environmental impact. This approach:

a. Is the most conservative and will result in many overpressure scenarios having the most severe consequences. Therefore, all such overpressure scenarios will need to be mitigated by applying the same risk-reduction factors. This approach will not allow for any distinction in IPLs between scenarios with the potential for very large pressure rises and those in which the pressure rise will be limited but still above the code limit.

b. Is strict in its view of following the code of construction and essentially means that a failure to achieve full code compliance is the consequence of concern.

c. Is easy to apply and enforce.

d. Reinforces the importance of adhering to design codes.

This option should be followed if there is no supporting data to demonstrate the integrity of the vessel.

Option 2: Assume that there is a hierarchy of consequences, based upon the overpressure that occurs. This approach:

a. Explicitly considers that, while a pressure rise above code allowance is undesirable, the consequence and likelihood of specific failure (i.e., leak versus rupture) depends upon the percent overpressure. See Table E-1 for a conceptual indication of the types of consequences that might be considered appropriate at different levels of overpressure.

b. Can allow consideration of a leak of a toxic material to be a consequential event, and yet distinguish that a small leak of a flammable material may be much less consequential.

c. Is more complex to apply and to enforce than Option 1.

d. Has the potential to lower the emphasis of maintaining operating parameters below prescriptive safety limits.

e. Places a greater dependence on the ITPM program to detect and correct defects that would reduce the safety margin of the vessel.

The potential consequences of various levels of overpressure should be defined by an individual organization, based on their design criteria, applicable codes, and the organization's risk protocol. An example hierarchy of possible consequences as a function of vessel overpressure is given in Table E-1.

Table E-1. Conceptual consequence vs. pressure vessel overpressure

(This example pertains to design of carbon steel vessels per ASME Boiler and Pressure Vessel Code (BPVC) Section VIII, Division 1 (2013); for other design codes and for other materials and grades, the consequence versus % accumulation may be more severe. Also, catastrophic failures at lower overpressures would be expected if the vessel is beyond its corrosion allowance, has experienced over temperature or overpressure excursions in the past, is operated below its ductile/brittle transition temperature, or exhibits pitting or cracking.)

Accumulation (% over MAWP)	Significance	Potential Consequence
10%	Allowable accumulation for process upset cases (non-fire) protected by a single relief device	No expected consequence at this accumulation level
16%	Allowable accumulation for process upset cases protected by multiple relief devices	No expected consequence at this accumulation level
21%	Allowable accumulation for external fire relief cases, regardless of the number of relief devices	No expected consequence at this accumulation level
>21% to 30%	Standard hydrostatic test pressure	Increasing likelihood of leaks in associated flanges, piping, equipment, etc.
>30%	Minimum yield and ultimate strength varies with material and grade	Catastrophic failure becomes increasingly likely. Since this level of overpressure goes beyond code allowance, an analysis and supporting documentation by the organization will be necessary to evaluate the severity of the

		consequence of overpressure.

The approach presented in Table E-1 is in general accordance (with the exception noted) with recommended changes of Code Case 2211-1 of the ASME Code presented in Welding Research Council Bulletin 498, *Guidance on the Application of Code Case 2211 – Overpressure by Systems Design* (Sims 2005). Also see UG-140 of Section VIII Division 1 of the ASME Boiler and Pressure Code (2013) for a better understanding of how this code case was incorporated into the pressure vessel code. Table E-1 only presents a conceptual example. Specific values for overpressure will need to be determined by individual organizations and account for materials of construction and design codes.

REFERENCES

ASME (American Society of Mechanical Engineers). 1999. ASME Boiler and Pressure Vessel Code Section VIII – *Rules for Construction of Pressure Vessels,* Division 1. New York: ASME.

ASME. 2013. ASME Boiler and Pressure Vessel Code Section VIII – *Rules for Construction of Pressure Vessels,* Division 1. New York: ASME.

Sims, R., and W. Yeich. 2005. *Guidance on the Application of Code Case 2211 – Overpressure Protection by Systems Design.* Bulletin 498. New York: Welding Research Council.

REFERENCES

ACGIH (American Council of Governmental Industrial Hygienists). 2004. *Industrial Ventilation: A Manual of Recommended Practice.* 25th Edition. Cincinnati: ACGIH.

ANSI/AIHA/ASSE (National Standards Institute/American Industrial Hygiene Association/American Society of Safety Engineers). 2012. *Fundamentals Governing the Design and Operation of Local Exhaust Ventilation* Systems. Z9.2. Falls Church, VA: AIHA.

ANSI/API (American National Standards Institute/American Petroleum Institute). 2008. *Guide for Pressure-Relieving and Depressuring Systems: Petroleum petrochemical and natural gas industries-Pressure relieving and depressuring systems,* 5th Edition addendum. (ISO 23251 Identical). API Standard 521. Washington, DC: API.

ANSI/ASME (American National Standards Institute/American Society of Mechanical Engineers). 2012. *Safety Standards for Conveyors and Related Equipment.* B20.1-2012. New York: ASME.

ANSI/CEMA (American National Standards Institute/Conveyor Equipment Manufacturers Association). 2008. *Screw Conveyors for Bulk Materials 350-2009.* Naples, FL: Conveyor Equipment Manufacturers Association.

ANSI/FCI (American National Standards Institute/Fluid Controls Institute). 2006. *Control Valve Seat Leakage,* ANSI/FCI 70-2-2006. Cleveland: Fluid Controls Institute, Inc.

ANSI/ISA (American National Standards Institute/International Society of Automation). 2004. *Functional Safety: Safety Instrumented Systems for the Process Industry Sector – Part 1: Framework, Definitions, System, Hardware and Software Requirements.* 84.00.01-2004 (IEC 61511-1 Mod). Research Triangle Park, NC: ISA.

ANSI/ISA. 2009. *Management of Alarm Systems for the Process Industries.* ANSI/ISA-18.2-2009. Research Triangle Park, NC: ISA.

ANSI/ISA. 2011. Wireless systems for industrial automation: Process controls and related applications. 100.11a-2011. Research Triangle Park, NC: ISA.

ANSI/ISA. 2012. *Identification and Mechanical Integrity of Safety Controls, Alarms and Interlocks in the Process Industry.* ANSI/ISA-84.91.01-2012. Research Triangle Park, NC: ANSI/ISA.

API (American Petroleum Institute). 1991. *Removal of Benzene from Refinery Wastewater.* Standard 221. Washington, DC: API.

API. 1996. *Evaluation of the Design Criteria for Storage Tanks with Frangible Roofs Joints.* Publication 937. Washington, DC: API.

API. 2002. *Pumps – Shaft Sealing Systems for Centrifugal and Rotary Pumps,* 2nd Edition. Standard 682. Washington, DC: API.

API. 2008. *Design and Construction of Large, Welded, Low-Pressure Storage Tanks,* 11th Edition. Standard 620. Washington, DC: API.

API. 2009a. *Tank Inspection, Repair, Alteration, and Reconstruction.* Standard 653. Washington, DC: API.

API. 2009b. *Valve Inspection and Testing.* Standard 598. Washington, DC: API.

API. 2009c. *Venting Atmospheric and Low-Pressure Storage Tanks,* 6th Edition. API 2000/ISO 28300 (Identical). Washington, DC: API.

API. 2013. *Welded Storage Tanks for Oil Storage,* 12th Edition. Standard 650. Washington, DC: API.

Arner, D., and H. Thomas. 2012. "Proven in Use (What's the Quality of Your Data?)." Paper presented at 8th Global Congress on Process Safety, Houston, TX, April 1-4.

ASME (American Society of Mechanical Engineers). 1999. ASME Boiler and Pressure Vessel Code Section VIII – *Rules for Construction of Pressure Vessels,* Division 1. New York: ASME.

ASME. 2013. ASME Boiler and Pressure Vessel Code Section VIII – *Rules for Construction of Pressure Vessels,* Division 1. New York: ASME.

ASTM (American Society for Testing and Materials). 2010. *Standard Test Method for Performance Testing of Excess Flow Valves.* Standard F1802-04. Washington, DC: ASTM International.

ATEX (Appareils destinés à être utilisés en Atmosphères Explosives). 2009. *Equipment for Explosive Atmospheres.* Directive 94/9/EC. Brussels: EC.

Bailey, R. 1997. "Estimation from Zero Failure Data." *Risk Analysis* Vol. 17. No. 3:375-380.

Baker, J., F. Bowman, G. Erwin, S. Gorton, D. Hendershot, N. Leveson, S. Priest, I. Rosenthal, P. Tebo, D. Wiegmann, and L. Wilson. 2007. "The Report of the BP US Refineries Independent Safety Review Panel; 2007." http://www.bp.com/liveassets/bp_internet/globalbp/globalbp_uk_english/SP/STAGING/local_assets/assets/pdfs/Baker_panel_report.pdf..

Bell, B., and A. Swain. 1983. *A Procedure for Conducting a Human Reliability Analysis for Nuclear Power Plants.* NUREG/CR-2254. Washington, DC: Nuclear Regulatory Commission.

Bento J.-P., S. Björe, G. Ericsson, A. Hasler, C.-D. Lydén, L. Wallin, K. Pörn, O. Akerlund. 1987. *Reliability Data Book for Components in Swedish Nuclear Power Plants.* RKS/SKI 85-25. Stockholm: Swedish Nuclear Power Inspectorate and Nuclear Training & Safety Center of the Swedish Utilities.

Blackman, H., D. Gertman, and R. Boring. 2008. "Human Error Quantification Using Performance Shaping Factors in the SPAR-H Method." Paper presented at Idaho National Labs, 52nd Annual Meeting of the Human Factors and Ergonomics Society, New York, NY, September 22-26.

Bridges, W. and T. Clark. 2011. "LOPA and Human Reliability – Human Errors and Human IPLs (Updated)." Paper presented at 2012 AIChE Spring Meeting/8th Global Congress on Process Safety, Houston, TX, April 1-5.

BS (British Standards Institution). 1984. *Specification for Valves for Cryogenic Service.* 6364. London: IBN.

BS. 2005. *Specification for the design and manufacture of site built, vertical, cylindrical, flat-bottomed, above ground, welded, steel tanks for the storage of liquids at ambient temperature and above.* EN 14015:2004 British-Adopted European Standard. London: IBN.

BS. 2006. *Dust explosion venting protective systems.* EN 14491. London: IBN.

Bukowski, J. and W. Goble. 2009. "Analysis of Pressure Relief Valve Proof Test Data." Paper presented at 2009 AIChE Spring Meeting/11th Plant Process Safety Symposium, Tampa, FL, April 26-30.

CCPS. 1989. *Guidelines for Process Equipment Reliability Data, with Data Tables.* New York: AIChE.

CCPS. 1993. *Guidelines for Safe Automation of Chemical Processes.* New York: AIChE.

CCPS. 1994. *Guidelines for Preventing Human Error in Process Safety.* New York: AIChE.

CCPS. 1998a. *Guidelines for Improving Plant Reliability through Data Collection and Analysis.* New York: AIChE.

CCPS. 1998b. *Guidelines for Pressure Relief and Effluent Handling Systems.* New York: AIChE.

CCPS. 1999. *Guidelines for Consequence Analysis of Chemical Releases.* New York: AIChE.

CCPS. 2000. *Guidelines for Chemical Process Quantitative Risk Analysis,* 2nd Edition. New York: AIChE.

CCPS. 2001. *Layer of Protection Analysis: Simplified Process Risk Assessment.* New York: AIChE.

CCPS. 2003. *Guidelines for Investigating Chemical Process Incidents,* 2nd Edition. New York: AIChE.

CCPS. 2006. *Guidelines for Mechanical Integrity Systems.* New York: AIChE.

CCPS. 2007a. *Guidelines for Risk Based Process Safety.* New York: AIChE.

CCPS. 2007b. *Guidelines for Safe and Reliable Instrumented Protective Systems.* New York: AIChE.

CCPS. 2007c. *Human Factors Methods for Improving Performance in the Process Industries.* New York: AIChE.

CCPS. 2008a. *Guidelines for Hazard Evaluation Procedures*, 3rd Edition. New York: AIChE.

CCPS. 2008b. *Guidelines for the Management of Change for Process Safety.* New York: AIChE.

CCPS. 2009a. *Guidelines for Developing Quantitative Safety Risk Criteria*, 2nd Edition. New York: AIChE.

CCPS. 2009b. *Inherently Safer Chemical Processes: A Life Cycle Approach,* 2nd Edition. New York: AIChE.

CCPS. 2013. *Guidelines for Enabling Conditions and Conditional Modifiers in Layer of Protection Analysis.* New York: AIChE.

CCPS. n.d. Process Equipment Reliability Database (PERD). http://www.aiche.org/ccps/resources/perd.

CDC (Centers for Disease Control and Prevention). 1998. *Controlling Cleaning-Solvent Vapors at Small Printers.* Number 98-107. Washington, DC: NIOSH.

Chlorine Institute. 2011. *Bulk Storage of Liquid Chlorine,* 8th Edition. Pamphlet 5. Arlington, VA: Chlorine Institute.

CPR (Committee for the Prevention of Disasters). 2005. *Guidelines for Quantitative Risk Assessment,* 2nd Edition "Purple Book." 18E. The Hague: Sdu Uitgevers.

Crane Co. 2009. *Flow of Fluids: Through Valves, Fittings and Pipe.* Technical Paper 410. Stamford, CT: Crane Company.

Cullen, W. (Lord). 1990. *The Public Inquiry into the Piper Alpha Disaster,* Volume 1. London: HMSO.

DIN (Deutsches Institut für Normung). 1997. *Plastics Piping Systems – Thermoplastic Valves – Test Methods for Internal Pressure and Leak-tightness.* EN 917. Berlin: Beuth Verlag BmbH.

DIN. 2006. *Explosion suppression systems.* EN 14373:2006. Berlin: Beuth Verlag BmbH.

Earles, D., and M. Eddins. 1962. *Failure Rates – Reliability Physics.* Proceedings of First Annual Symposium on the Physics of Failure in Electronics. Baltimore: Spartan Books.

EC (European Commission). 1997. *Pressure Equipment Directive 97/23/EC* (PED). Brussels: EC.

Edwards, E., and F. Lees. 1974. *The Human Operator in Process Control.* London: Taylor & Francis Ltd.

EGIG (European Gas pipeline Incident data Group). 2011. *Gas Pipeline Incidents, 8th Report of the European Gas Pipeline Incident Data Group.* Groningen: EGIG.

Embrey, D., P. Humphreys, E. Rosa, B. Kirwan, & K. Rea. 1984. *SLIM-MAUD: An approach to assessing human error probabilities using structured expert judgement.* NUREG/CR-3518. Washington DC: U.S. Nuclear Regulatory Commission.

exida. 2006. *Electrical & Mechanical Component Handbook.* Sellersville, PA: exida.

Exida. 2007. *Safety Equipment Reliability Handbook,* 3rd Edition. Sellersville, PA: Exida.

Freeman, R. 2011. "What to Do When Nothing Has Happened," *Process Safety Progress,* September, Vol. 30, No. 3: 204-211.

Freeman, R., and D. Shaw. 1988. "Sizing Excess Flow Valves," *Plant/Operations Progress,* July, Vol. 7, No. 3:176-182.

Gertman, D., and H. Blackman. 1994. *Human Reliability and Safety Analysis Data Handbook.* New York: John Wiley & Sons.

Gertman, D., H. Blackman, J. Marble, J. Byers, and C. Smith. 2005. *The SPAR-H Human Reliability Analysis Method.* NUREG CR-6883. Washington, DC: U.S. Nuclear Regulatory Commission, Office of Nuclear Regulatory Research.

Goodrich, M. 2010. "Risk Based Pump-Seal Selection Guideline Complementing ISO21049/API 682." Proceedings of the Twenty-Sixth International Pump Users Symposium, Texas A&M University. http://turbolab.tamu.edu/articles/26th_international_pump_users_symposium_proceedings.

Grossel, S. 2002. *Deflagration and Detonation Flame Arresters.* New York: CCPS/AIChE.

Grossel, S. and R. Zalosh. 2005. *Guidelines for Safe Handling of Powders and Bulk Solids.* New York: CCPS/AIChE.

Hallbert, B., A. Whaley, R. Boring, P. McCabe, and Y. Chang. 2007. *Human Event Repository and Analysis (HERA): The HERA Coding Manual and Quality Assurance* (NUREG CR-6903, Volume 2). Washington, DC: Division of Risk Assessment and Special Projects, Office of Nuclear Regulatory Research, U.S. Nuclear Regulatory Commission.

Hannaman, G., A. Spurgin, and Y. Lukic. 1984. *Human cognitive reliability model for PRA analysis.* Draft Report NUS-4531, EPRI Project RP2170-3. Palo Alto: Electric Power and Research Institute.

IEC (International Electrotechnical Commission). 2003. *Functional Safety: Safety Instrumented Systems for the Process Industry Sector – Part 1: Framework, Definitions, System, Hardware and Software Requirements.* IEC 61511. Geneva: IEC.

IEC. 2004. *Reliability data handbook – Universal model for reliability prediction of electronics components, PCBs and equipment.* IEC 62380. Geneva: IEC.

IEC. 2010. *Functional safety of electrical/electronic/programmable electronic safety related systems.* IEC 61508. Geneva: IEC.

IEEE (Institute of Electrical and Electronics Engineers). 1984. *Guide to the Collection and Presentation of Electrical, Electronic, Sensing Component, and Mechanical Equipment Reliability Data for Nuclear-Power Generation Stations.* Standard 500-1984 reaffirmed 1991. New York: IEEE.

IEEE. 2007. *Recommended Practice for the Design of Reliable Industrial and Commercial Power Systems.* Standard 493-2007. New York: IEEE.

ISA (International Society of Automation). 2002. *Safety Instrumented Functions (SIF) – Safety Integrity Level (SIL) Evaluation Techniques.* TR84.00.02-

2002. Research Triangle Park, NC: ISA.

ISA. 2011. *Guidelines for the Implementation of ANSI/ISA 84.00.01.* TR84.00.04-2011 Part 1. Research Triangle Park, NC: ISA.

ISA. 2012. *Mechanical Integrity of Safety Instrumented Systems (SIS).* TR84.00.03-2012. Research Triangle Park, NC: ISA.

ISO (International Organization for Standardization). 1985. *Explosion protection systems -- Part 4: Determination of efficacy of explosion suppression systems.* 6184-4:1985. Geneva: ISO.

Kirwan, B. 1994. *A Guide to Practical Human Reliability Assessment.* Boca Raton, FL: CRC Press.

Linstone, H. 1975. *The Dephi Method.* Boston: Addison-Wesley.

Lorenzo, D. 1990. *A Manager's Guide to Reducing Human Error: Improving Human Performance in the Chemical Industry.* Washington, DC: Chemical Manufacturers Association.

M2K (Maintenance 2000). 2000. *FARADIP – Electronic, electrical, mechanical, pneumatic equipment.* [computer program] http://www.m2k.com/failure-rate-data-in-perspective.htm.

Meyer, M., and J. Booker. 2001. *Eliciting and Analyzing Expert Judgment.* Philadelphia: Society for Industrial and Applied Mathematics.

MSS (Manufacturers Standardization Society). 2009. *Pressure Testing of Valves.* MSS-SP-61. Vienna, VA: MSS.

Mudan, K. S. 1984. "Thermal Radiation Hazards from Hydrocarbon Pool Fires." *Progress in Energy Combustion Science.* 10:59-81.

NFPA (National Fire Protection Association). 2007. *Standard on Explosion Protection by Deflagration Venting.* 68. Quincy, MA: NFPA.

NFPA. 2008a. *Flammable and Combustible Liquids Code.* 30. Quincy, MA: NFPA.

NFPA. 2008b. *Standard for the Inspection, Testing, and Maintenance of Water-Based Fire Protection Systems.* 25. Quincy, MA: NFPA.

NFPA. 2008c. *Standard on Explosion Prevention Systems.* 69. Quincy, MA: NFPA.

NFPA. 2009. *Standard for Dry Chemical Extinguishing Systems.* 17. Quincy, MA: NFPA.

NFPA. 2010a. *Standard for Exhaust Systems for Air Conveying of Vapors, Gases, Mists, and Noncombustible Particulate Solids.* 91. Quincy, MA: NFPA.

NFPA. 2010b. *Standard for Low-, Medium-, and High-Expansion Foam.* 11. Quincy, MA: NFPA.

NFPA. 2012. *Standard on Clean Agent Fire Extinguishing Systems.* 2001. Quincy, MA: NFPA.

NFPA. 2013. *Standard for the Prevention of Fire and Dust Explosions from the Manufacturing, Processing, and Handling of Combustible Particulate Solids.* 654. Quincy, MA: NFPA.

NOAA/U.S. EPA (National Oceanic and Atmospheric Administration/U.S. Environmental Protection Agency). 2012. *ALOHA (Areal Locations of Hazardous Atmospheres)* (Version 5.4.3) [computer program] http://www.epa.gov/emergencies/content/cameo/aloha.htm.

NRCC (National Research Council Canada). 1995. *Water Mains Breaks Data on Different Pipe Materials for 1992 and 1993*, A-7019.1. Ontario: National Research Council.

NTSB (National Transportation Safety Board). 1990. *Marine Accident Report: The Grounding of the U.S. Tankship Exxon Valdez on Bligh Reef, Prince William Sound Near Valdez, Alaska March 24, 1989*, Washington, DC: NTSB.

OREDA. 2009. *Offshore Reliability Data Handbook*, 5th Edition *(OREDA)*. Trondheim, Norway: SINTEF.

Rasmussen, J., and W. Rouse. 1981. *Human Detection and Diagnosis of System Failures*. New York: Plenum Press.

Reason, J. 1990. *Human Error*. New York: Cambridge University Press.

Reason, J. and A. Hobbs. 2003. *Managing Maintenance Error: A Practical Guide*. Hampshire, U.K.: Ashgate Publishing Company.

RIAC (Reliability Information Analysis Center). 1991. *Failure Mode/Mechanism Distributions*. FMD-91. Rome, NY: RAC.

RIAC. 1995. *Nonelectronic Parts Reliability Database*. NPRD-95. Rome, NY: RAC.

RIAC. 1997. *Electronic Parts Reliability Data (EPRD)*. NPRD-97. Rome, NY: RAC.

Sims, R., and W. Yeich. 2005. *Guidance on the Application of Code Case 2211 – Overpressure Protection By Systems Design*. Bulletin 498. New York: Welding Research Council.

SINTEF (Stiftelsen for industriell og teknisk forskning). 2010. *PDS Data Handbook*. Trondheim, Norway: SINTEF.

SIS-TECH. 2012. *SIL Solver®*. (Version 6.0) [computer program] http://sis-tech.com/software.

Spirax Sarco. 2013. "Steam Engine Tutorials," Tutorial 12, "Piping Ancilliaries," http://www.spiraxsarco.com/resources/steam-engineering-tutorials.asp.

SRC (Alion System Reliability Center). 2006. SPIDR™ – System and Part Integrated Data Resource [computer program] http://src.alionscience.com/spidr/.

Summers, A. 2014. "Safety Controls, Alarms, and Interlocks as IPLs." *Process Safety Progress*, Vol. 33, No. 2, June 2014, p. 184-194.

Swain, A., and H. Guttmann. 1983. *Handbook of Human Reliability Analysis with Emphasis on Nuclear Power Plant Applications*. NUREG CR-1278. Washington, DC: U.S. Nuclear Regulatory Commission.

Telcordia Technologies. 2011. *Reliability Prediction for Electronic Equipment*. SR-332. Morristown,NJ: Telcordia.

UL (Underwriters Laboratories). 2007. *Steel Aboveground Tanks for Flammable and Combustible Liquids*. UL-142. Northbrook, IL: UL.

UNM (Union de Normalisation de la Mécanique). 2002. *Unfired Pressure Vessels.* EN 13445. Courbevoie: UNM.

USCG (United States Coast Guard). 2011. *Report of Investigation into the Circumstances Surrounding the Explosion, Fire, Sinking and Loss of Eleven Crew Members Aboard the Mobile Offshore Drilling Unit DEEPWATER HORIZON in the Gulf of Mexico, April 20-22, 2010* (Volume 1). Washington, DC: USCG.

U.S. CSB (Chemical Safety and Hazard Investigation Board). 2007. *Emergency Shutdown Systems for Chlorine Transfer.* Safety Bulletin No. 2005-06-I-LA. Washington, DC: CSB.

U.S. DoD (Department of Defense). 1990. *Reliability Prediction of Electronic Equipment.* Military Handbook 217F. Washington, DC: U.S. Department of Defense.

U.S. EPA (Environmental Protection Agency). 2003. *Cross Connection Control Manual.* EPA 816-R-03-002. Washington, DC: EPA.

U.S. EPA. 2007. *Emergency Isolation for Hazardous Material Fluid Transfer Systems – Application and Limitations of Excess Flow Valves.* Chemical Safety Alert, EPA 550-F-0-7001. Washington, DC: EPA.

U.S. EPA. 2009. ECA Workshop for Alabama Correction Facilities, SPCC Review. Washington, DC: EPA.

U.S. NRC (Nuclear Regulatory Commission). 2008. *Fitness for Duty Programs – Managing Fatigue.* Regulation 10 CFR 26 Subpart I. Washington, DC: NRC.

U.S. OSHA (Occupational Safety & Health Administration). 2005 *Flammable Liquids.* 29 CFR 1910.106. 1974-2005. Washington, DC: OSHA.

VDI (Verein Deutscher Ingenieure). 2002. *Pressure Venting of Dust Explosions.* 3673. Berlin: VDI.

Walpole, R., R. Myers, S. Myers, and K. Ye. 2006. *Probability and Statistics for Engineers and Scientists,* 8th Edition. Upper Saddle River, NJ: Pearson Publishing.

Welker, E., and M. Lipow. 1974. *Estimating the Exponential Failure Rate from Data with No Failure Events.* Proceedings of the 1974 Annual Reliability and Maintainability Symposium, Los Angeles, California. New York: IEEE.

INDEX

Printed and bound by CPI Group (UK) Ltd, Croydon, CR0 4YY

23/04/2025

14660911-0001